Centrifugal Pump Sourcebook

John W. Dufour

W. Ed Nelson

McGraw-Hill, Inc.

New York St. Louis San Francisco Auckland Bogotá
Caracas Lisbon London Madrid
Mexico Milan Montreal New Delhi Paris
San Juan São Paulo Singapore
Sydney Tokyo Toronto

Library of Congress Cataloging-in-Publication Data

Dufour, John W.
  Centrifugal pump sourcebook / John W. Dufour, W. Ed Nelson.

      p.   cm.
  Includes index
  ISBN 0-07-018033-4
  1. Centrifugal pumps—Handbooks, manuals, etc. I. Nelson, W. Ed.
II. Title.
TJ919.D84   1992   92-9989
621.6'7dc20   CIP

1 2 3 4 5 6 7 8 9 0   DOC/DOC   9 7 6 5 4 3 2

ISBN 0-07-018033-4

*The sponsoring editor for this book was Robert W. Hauserman, the
editing supervisor was Jim Halston, and the production supervisor
was Suzanne W. Babeuf. It was set in Century Schoolbook by The
Clarinda Company.*

Printed and bound by R. R. Donnelley & Sons Company.

Contents

Chapter 12. Pump System Troubleshooting

Preface

Many different types of pumps are used in industry, but the most common is the centrifugal pump. This book deals with the basic design, application, and maintenance of centrifugal pumps. The book is intended to be not just another pump handbook, but a sourcebook of practical information about centrifugal pumps that the authors have acquired over many years of experience. The book is designed to be used as both a quick reminder guide for the experienced rotating equipment engineer and a detailed reference book for the novice field maintenance engineer.

The first two chapters cover pump discharge performance. Chapter 1 discusses centrifugal pump technology, including such topics as centrifugal pump functions, basic pump hydraulics, and impeller design. Pump curve versus system curve, creating a system sketch, and calculating $NPSH_a$ are some of the areas covered in Chapter 2. The equally important area of pump suction performance is addressed in Chapter 3 in sections on NPSH, vortexing, recirculation, and minimum flow requirements.

Radial hydraulic loads, corrective actions, radial clearances, shaft failures, and axial thrust problems are among the topics presented in Chapter 4.

Chapters 5 and 6 discuss various types of bearings including anti-friction bearings, ball bearings, and journal or sleeve-type bearings. Chapter 7 covers shaft sealing, packing, mechanical seals, and seal flush systems.

A number of mechanical considerations, including lateral critical speeds, dynamic balancing of components, process pump gaskets, and impeller attachment methods, are presented in Chapter 8.

Vertical pumps are being used more and more in industry because their operational features are advantageous in many situations. Chapter 9 is devoted to vertical pumps, covering such areas as vertical pump theory, construction details, applications, vortexing, and specifications.

Chapters 10 and 11 discuss installation and alignment techniques, respectively. Chapter 10 gives general procedures for centrifugal pump installation as well as information on grouting and mounting. Alignment, couplings, and dial indicators are some of the topics covered in Chapter 11.

Chapter 12 addresses the issue of pump system troubleshooting. Components of a pumping system, vibration-monitoring basics, and power transmission vibrations are just a few of the items presented in this chapter.

John W. Dufour
W. Ed Nelson

Centrifugal
Pump
Sourcebook

Introduction

The basic principles of pumps and the differences between centrifugal and positive displacement types determine the applications for which each style is best suited. The term *centrifugal pump* over the years has become a generic term used to describe a wide range of pumping designs. Generally, this type of pump has one or more impellers equipped with vanes, which are rotated by a driver (such as an electric motor or steam turbine), within a closely fitted casing. Within the pump, energy is transferred from the rotating impeller to the liquid passing through it, thus raising the kinetic energy. Subsequently, the volute or diffuser section of the casing converts this energy to potential energy, raising the discharge pressure. The efficiency at which this energy transfer takes place—and therefore the change in the discharge pressure—varies with changes in flow rate through the pump.

The history of the centrifugal pump can be traced to the late 1600s. However, the modern centrifugal pump, as we now know it, came into general use only within the past hundred years. The progress of its development paralleled that of the electric motor and modern steam turbine; as these drivers became bigger and more powerful, so did the centrifugal pump. The early years of the pump industry were characterized by a large number of small manufacturers, frequently family-owned businesses. They established themselves by designing and building pumps to meet specific needs in the marketplace. The general design and layout of the centrifugal pump have not changed much in the last fifty years. However, improvements in containment, efficiency, and reliability have been made and continue to be made. The main focus in the next few years will be on cost reduction, reduced emissions from sealing devices, and continued improvements in performance and reliability. These developments will most likely include the use of new and innovative pump controls, such as variable-speed motors and telemetry systems; the application of modern materials, such as composites, ceramics, and plastics; and sealless pumps, possibly with canned rotors and magnetic bearings. In addition, with the aid of computers and modern flow visualization techniques, new impeller and casing configurations are rapidly being designed.

Today the pump marketplace is still very complex. There exist a large number and variety of pump manufacturers, with the majority still being relatively small companies serving a specialty market. There are only a few large, diversified companies that are capable of producing a wide spectrum of pump products for the industries they serve, including municipal water and waste, pulp and paper, conventional and nuclear power, refining and petrochemicals, as well as the military and space programs.

The worldwide oil refining and petrochemical industry is one of the largest users of centrifugal pumps. Applications include fluids that can be (and often are) both flammable and toxic. Pumping parameters typically have temperatures from -50 to over 700°F, pressures from 28 inches of mercury (inHg) vacuum to over 3000 pounds per square inch (lb/in^2) gage, and specific gravities from as low as 0.62 to over 2.0. This wide range of pumping conditions presents some unique challenges to both the manufacturer and the end user. This book deals primarily with centrifugal pumps designed for the petroleum industry. However, the knowledge gained, the technology developed, and the maintenance techniques used in addressing the problems and needs of this specific industry are applicable to the majority of all centrifugal pumps.

The hydraulic performance of a centrifugal pump in the field can differ significantly from the published performance curve. The potential causes of this problem may involve the hydraulic design of the pump, while at other times the reliability of the pump may be compromised as a result of the failure of or because of improperly manufactured or assembled mechanical parts. Frequently, to determine the cause of a problem, the pump must be taken out of service and dismantled for inspection, and each major component must undergo careful dimensional analysis. The causes of field problems are often difficult to identify and to correct. However, to properly diagnose a centrifugal pump problem, the maintenance engineer must always remember that the hydraulic design and the mechanical construction of a centrifugal pump are inextricably interrelated.

Most traditional pump handbooks consist of many pages of technical data, pressure-drop tables, conversion charts, etc., but they fail to tie the whole pump together. Little explanation is given of the practical considerations of centrifugal pump selection, operation, and maintenance. The following three brief paragraphs, not normally found in most pump handbooks, help to show the relationship between hydraulic conditions and possible pump mechanical problems:

1. Centrifugal pumps should be selected and normally operated at or near the manufacturer's design rated conditions of head and flow. This is usually at the *best efficiency point* (BEP). Pump impeller vane angles

and the size and shape of the internal liquid flow passages are fixed and can only be designed for one point of optimum operation. For any other flow conditions, these angles and liquid channels are either too large or too small.

2. Any pump operated at excess capacity, i.e., at a flow significantly greater than BEP and at a lower head, will surge and vibrate, creating potential bearing and shaft seal problems as well as requiring excessive power.

3. When operation is at reduced capacity, i.e., at a flow significantly less than BEP and at a higher head, the fixed vane angles will now cause eddy flows within the impeller, casing, and between the wear rings. The radial thrust on the rotor will increase, causing higher shaft stresses, increased shaft deflection, and potential bearing and mechanical seal problems while radial vibration and shaft axial movement will also increase. Continued operation in this mode will result in the accelerated deterioration of the mechanical and hydraulic performance and may ultimately result in the failure of the pump.

Centrifugal Pump Technology

1.1 Centrifugal Pump Functions

Centrifugal pumps are applied to many different types of liquid services. However, these applications can generally be categorized into three basic functions or combinations of functions (Fig. 1.1a to c):

- To raise a liquid from one level to another, i.e., from a below-grade sump to an elevated holding tank.

- To force a liquid into a vessel at a higher pressure by overcoming internal pressure; i.e., a boiler water feed pump takes suction from a storage tank operating below atmospheric pressure and discharges to a high-pressure boiler.

- To force a liquid through a pipe by overcoming piping and fitting friction losses; i.e., a petroleum pipeline pump is used to move millions of gallons of crude oil and refined product across the country through miles of underground pipe.

1.2 Basic Pump Hydraulics

The first law of thermodynamics is the theoretical basis of all centrifugal pumps. Daniel Bernoulli first presented his energy theorem in the early 1730s; it was to become one of the most important concepts in fluid dynamics in that it expresses the concept of the conservation of energy:

> The internal or mechanical energy of a system appears as kinetic, pressure, and potential energy; these are *interchangeable* and *their sum must remain constant.*

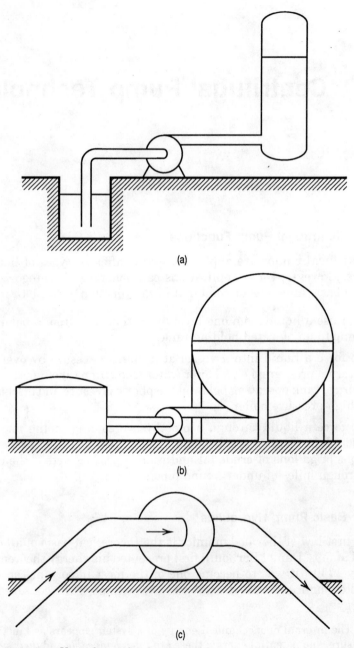

Figure 1.1 Uses of centrifugal pumps.

Stated another way, the law says that

> . . . in an ideal system, energy can be neither created nor destroyed but can only be *transformed* from one form to another.

The fluid entering a centrifugal pump has some initial energy. That energy is increased within the pump as work is done on the fluid. Actually, the pump transforms mechanical energy from the rotating driver (kinetic energy) to pressure or head (potential energy) of the discharged liquid.

Head is a term that expresses the energy of a fluid due to its elevation above some reference point (such as a pump centerline), its velocity, and its pressure. Head, usually expressed in feet of water, is very important in understanding and describing centrifugal pump design and performance. There are two fundamental pressure conditions of fluids: the static pressure or head, where fluids are at rest, and the dynamic head or pressure, where fluids are flowing. *Static head* is the height from a given point in a column or body of liquid at rest to the surface of the liquid, expressed in feet. The pressure at the reference point can be calculated by dividing the static head by 2.31. The factor 2.31 is the height of a column of water, in feet (ft), that will support 1 pound per square inch (lb/in^2) at its base. *Dynamic head* is a measure of the energy of motion of a flowing fluid. It is, in effect, the static head required to accelerate the liquid to its flowing velocity. In Fig. 1.2, the distance from the top of the water tank to the ground is the static head.

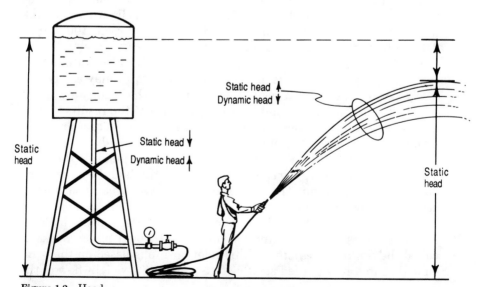

Figure 1.2   Head.

As the liquid flows down through the pipe, it loses static head but gains velocity, or dynamic head. At the point of discharge, the dynamic heads begins converting back to static head as the liquid rises, until all dynamic head is converted to static head and then falls back to the ground. The "loss of head" represents friction losses in the pipe and air. The relationship between the static head and the pressure for any liquid is determined as follows:

$$\text{Pressure (lb/in}^2) = \frac{(\text{static head, ft})(\text{SG})}{2.31} \tag{1.1}$$

where the specific gravity (SG) is the ratio of the weight of a volume, such as 1 cubic foot (ft^3), of any liquid to the weight of that same volume of water at 60°F.

You must be sure to use the specific-gravity factor when converting head to pounds per square inch and vice versa. Note that the total head in feet developed by a pump is the same regardless of the liquid pumped, whereas the value in pounds per square inch developed will vary directly with the specific gravity of the fluid. Therefore, the conversion of all pressure energy terms to units of equivalent head greatly simplifies most pump calculations.

For example, if the liquid in the tank is water (SG = 1.0) and it has a static height equal to 100 ft above the gage, the gage will read $(100 \times 1.0)/2.31 = 43.3$ lb/in² gage. If the liquid in the tank is replaced with another of equal volume but different specific gravity, then the level within the tank will remain the same but the pressure measured at the bottom will have changed. For example, if the liquid in the tank were kerosene (SG = 0.7), the pressure gage would read $(100 \times 0.7)/2.31 = 30.3$ lb/in² gage.

Note: Head is used to describe pump performance also out of a concern for safety. By using head, a pump manufacturer may easily and safely performance-test a centrifugal pump in the factory by using water. The manufacturer and the buyer know that the results will be directly applicable for most other fluids, including those that are flammable or toxic, which the pump may handle in the field.

The *differential head* developed by a centrifugal pump is the measure of the total energy *increase* imparted to the liquid by the work done within the pump. The *total head* is the discharge head *plus or minus* the suction head.

Another useful relationship, frequently used in the study of hydraulics, that helps in understanding how a centrifugal pump works is called the *continuity equation*. This relationship is based on the principle that in a liquid-filled steady-state system, the weight flow of the fluid, per unit of time, passing any point is constant. Since most liquids are practically incompressible (weight flow and volume flow rates are both constant), this concept may be put in equation form as follows:

$$V = \frac{0.321Q}{a} \qquad (1.2)$$

where V = velocity, feet per second (ft/s)

Q = volume flow, gallons per minute (gal/min)

a = area of pipe, in^2

Therefore, for a constant-volume flow rate through a pipe or a pump flow passage, the liquid's velocity (or kinetic energy) will *decrease* if the cross-sectional area of the pipe or pump casing is *increased* (Fig. 1.3). And from the first law of thermodynamics, if the fluid's kinetic energy *decreases,* the potential or pressure energy (head) must *increase.* This is exactly how an ideal centrifugal pump works. In a real pump there are losses (usually in the form of heat), but that is covered in greater detail later.

1.3 Nomenclature

The basic components of a centrifugal pump are shown in Fig. 1.4. The impeller is the heart of a centrifugal pump. Securely attached to a shaft, the impeller transmits rotational energy (velocity) from the driver to the fluid being pumped. The impeller (impellers for multistage pumps) is surrounded by a housing known as the casing which

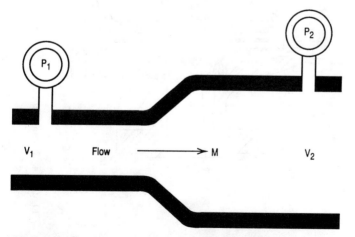

Figure 1.3 Continuity equation. Mass flow = constant; $P_1 < P_2$; $V_1 > V_2$.

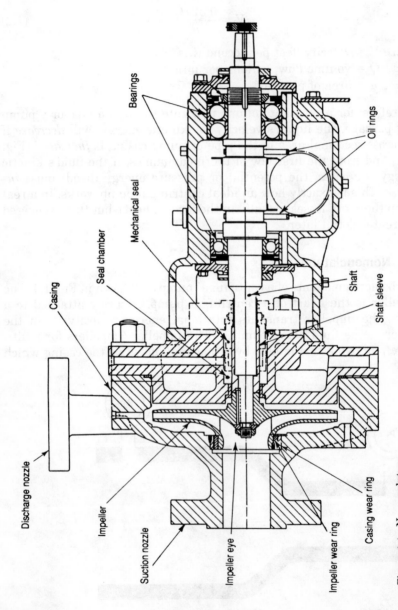

Figure 1.4 Nomenclature.

also supports the suction inlet and discharge outlet nozzles. The suction nozzle directs the fluid to the center or eye of the impeller. Once the liquid is caught up by the impeller vanes, it is accelerated to a higher velocity, thus increasing its kinetic energy. The maximum fluid kinetic energy (Fig. 1.5) is attained at the outlet of the vane, where its velocity approaches that of the tip velocity of the impeller. The exiting high-velocity fluid is contained in the surrounding pump casing and is directed toward the volute or diffuser section of the pump. Note how the cross-sectional area of the volute section increases from the entrance to the discharge nozzle. There within the volute, in accordance with the continuity equation, as the cross-sectional area increases, the fluid velocity decreases and as the fluid's kinetic energy decreases, its potential energy or pressure is increased.

To prevent any of the fluid that is being pumped from leaking out of the pump along the rotating shaft, a shaft sealing mechanism is used. This is normally a mechanical sealing device or a series of packing rings. Mechanical seals and packing are covered in more detail in Chap. 7.

To prevent higher-pressure fluid from leaking internally back to the suction side of the impeller, pairs of wear rings (one stationary and one rotating) having close clearances are used. Wear rings are covered in Chap. 8.

1.4 Hydraulic Design

Centrifugal pumps can be further classified by their housing or casing configuration. The most common type of centrifugal pump construction is the *volute* style (Fig. 1.5). A volute casing is a spiral or snaillike form designed such that as the liquid exits from the impeller into the casing, the cross-sectional area (volume) increases at a rate proportional to the increasing volume of liquid. Thus, in the ideal pump at the design flow rate, the fluid velocity within the spiral casing remains constant around the entire periphery of the impeller. This velocity is then converted to pressure in the volute or diffuser portion of the pump.

Another common centrifugal pump design is called the *vane* or *diffuser* style (Fig. 1.6). In this design, each set of diffuser vanes actually forms several small volutes or diverging passages in which the fluid velocity is converted to pressure. For multistage pumps, these diffusers also direct the fluid flow into the return passages, which in turn direct the fluid into the eye of the next stage impeller or discharge nozzle.

In addition to pump casing designs, centrifugal pumps may be clas-

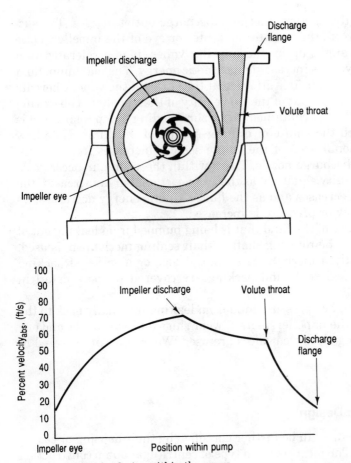

Figure 1.5 Fluid velocity within the pump.

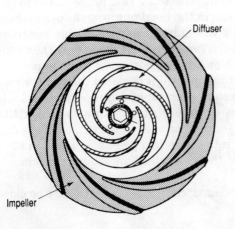

Figure 1.6 Diffuser pump.

sified by their impeller styles. Three basic impeller configurations are commonly used (Fig. 1.7):

- *Fully open:*   The vanes extend radially from the impeller hub without a back shroud or front cover plate.

- *Semiopen:*   The vanes project forward from the back shroud, and there is no cover plate.

- *Fully closed:*   A full back shroud and full front cover plate enclose and support the vanes.

Open and semiopen impellers are typically used on small centrifugal pumps where the liquids being pumped have a high solids content such as in sump, sludge separation, and slurry service.

Open and semiopen impellers allow larger particles to pass through the impeller that would normally become lodged in a fully closed design. However, due to internal leakage and recirculation, these impellers are generally less efficient than the fully closed designs. Therefore, in larger pumps, where the area between the vanes is larger, fully closed impellers are normally used in these services. The fully closed or

(a)

(b)

(c)

Figure 1.7   Impeller styles:
(*a*) fully open; (*b*) semiopen;
(*c*) fully closed.

just closed impeller is the most common type found in industry for both single-stage and multistage pump configurations.

1.5 Specific Speed

Pump designers use *discharge specific speed N_s*, a nondimensional design index, to classify pump impellers as to type and physical proportions. It is defined as the speed, in revolutions per minute (r/min), at which a geometrically similar impeller would operate if it were of such size as to deliver 1 gal/min against 1 ft of head. Specific speed is used by the pump designer to describe the geometric similarity between impellers (Fig. 1.8) and by the pump user to help understand the limitations of various impeller designs and how different impeller configurations affect the performance curve. This information is useful both during the initial pump selection and in rerating pumps and field troubleshooting.

The *specific speed* is defined as

$$N_s = \frac{(\text{speed, r/min})(\text{flow, gal/min})^{1/2}}{(\text{total head, ft})^{3/4}} \qquad (1.3)$$

When specific speed is calculated, the values of rotational speed, flow rate, and total head are *always* taken at the best efficiency point (BEP) with the maximum impeller diameter and rated speed.

The specific speed, impeller style, and pump configuration necessary to perform in a given set of flow conditions can be determined. As the

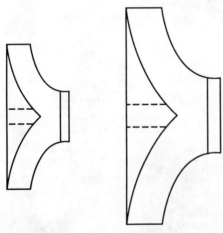

3600 r/min
300 gal/min
100 ft

1800 r/min
3000 gal/min
184 ft

Figure 1.8 Discharge specific speed example.

specific speed increases, the ratio of the impeller outlet diameter D_2 to the inlet or eye diameter D_1 decreases. This ratio becomes 1.0 for a true axial flow impeller. The specific speed determines the general shape or class of the impeller, as depicted in Fig. 1.9.

Impeller styles for a specific speed of 500 to 1000 are characterized by their production of high heads at low flows and by their plain or pure radial vanes with little or no curvature from the inlet eye to the outside diameter. Liquid flow through the impeller is primarily radial to the shaft.

The most common impeller styles have specific speeds from 1500 to 3000 and are sometimes called Francis-type impellers. These impellers

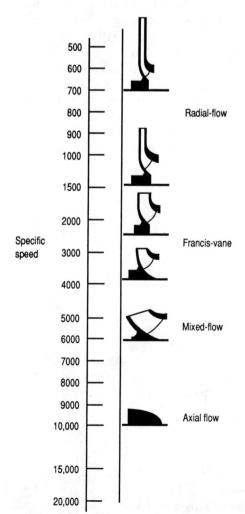

Figure 1.9 Specific speed versus impeller type.

produce medium heads at a wide range of flows and usually have vanes with a double curvature or twist. The term *pickup* is sometimes applied to impellers with Francis-type vanes. Because this type of vane extends into the eye of the impeller, it picks up the liquid before it makes the right-angle turn at the hub of the impeller. This transfers some of the work from the suction to the discharge side of the impeller and decreases pickup and entrance losses and undue shock. For this reason, a Francis-type impeller is frequently used for the first or suction stage of multistage pumps, and the radial-type impeller is used for the remaining stages.

Mixed-flow and axial-flow impeller designs have specific speeds ranging from 5000 to 20,000 (Fig. 1.9). These impellers are used for low-head high-volume applications such as in cooling water service. The vanes may have some curvature, but the flow through the impeller is primarily axial or parallel to the axis of the shaft.

Specific speed becomes a useful factor in evaluating both cost and efficiency for a given head and capacity. Figure 1.10 is a plot of efficiency versus specific speed for a set of typical, commercially available impellers from one manufacturer. Mechanical losses tend to remain relatively constant as the specific speed increases. Internal leakage losses and losses due to disk friction are more critical at low specific

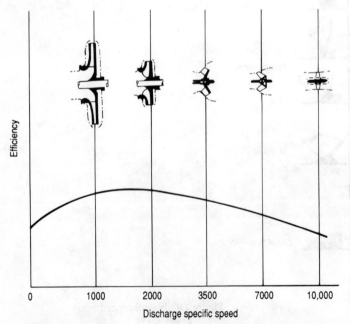

Figure 1.10 Specific speed versus efficiency for a given head and capacity.

speeds while hydraulic losses become a major factor at higher specific speeds. Therefore, efficiency peaks in the area where the specific speed is 2000 to 3000. In formula (1.3), we saw that specific speed varies directly as the pump rotational speed. Therefore, by selecting a pump speed which results in a specific speed near 2000, pump efficiency can usually be optimized. However, for a given capacity and head, the more rapidly the pump can be operated, usually the smaller the pump will be physically. Small, high-speed pumps as well as large, slow axial flow pumps tend to be less expensive than centrifugal pumps with specific speeds of 1000 to 3000, but the larger the pump, generally the higher the optimum efficiency.

1.6 Impeller Design: The Cause of the Pump Curve

1.6.1 Head-flow performance curve

For the ideal pump, having neither hydraulic shock nor internal friction losses, the head-flow curve is a sloping straight line. However, in the real world with real pumps, there are losses and the curve is not straight. Figure 1.11 is a typical head-flow (*H-Q*) centrifugal pump performance curve. Not all centrifugal pump curves look like this one. Some are flatter, others have humps, still others have wide flow ranges while some rise steeply from runout to shutoff over very small flow changes.

Friction losses increase as the flow increases. In addition,

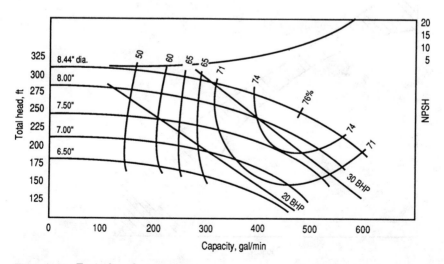

Figure 1.11 Typical performance curve.

most centrifugal pumps are manufactured from metal and have a fixed geometry. Pump engineers design that geometry so that all the flow areas, vane exit and inlet angles, impeller diameter, etc., are optimized to produce a given head at a given flow with minimal losses. This design point is called the *best efficiency point* (BEP). Operation on either side of the best efficiency flow point will cause hydraulic shock losses to develop. Hydraulic shock losses are more prevalent at low flows while friction losses are more of a problem at high flows. Together (Fig. 1.12) they combine to generate the actual pump curve which, depending on the pump design and configuration, can deviate significantly from the ideal, straight-line performance curve. The two major factors affecting the shape of the pump curve are the specific speed and impeller-casing interactions.

The pump curve can tell a lot about the physical characteristics of the impeller (or the impeller can tell a lot about the curve). Several simple statements can be made about the impeller pattern or style and the head-capacity (*H-Q*) or pump performance curve.

1. *Impeller width:*   Changing the width of the impeller of a centrifugal pump will usually alter the head-capacity curve. A wider impeller discharges a greater volume of liquid and generally has a flatter head-flow (*H-Q*) curve than a narrow one does. The pump

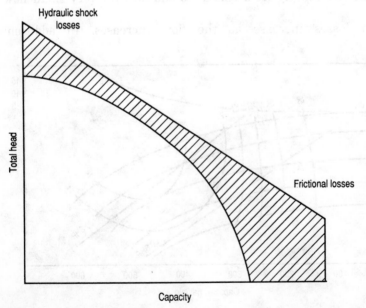

Figure 1.12   Cause of the curve.

with the narrower impeller generally has less capacity and a steeper *H-Q* curve.

2. *Number of vanes:* Increasing the number of vanes (10 maximum) in an impeller produces a flatter *H-Q* curve. Reducing the number of vanes (3 minimum) gives a steeper curve.

3. *Vane exit angle:* Changing the pitch of the impeller vanes alters performance. A radial or spokelike vane usually has a flatter *H-Q* curve. Sweptback vanes give a steeper *H-Q* curve. Exit angles normally vary from about 15 to 40°.

4. *Warp or sweepback of vanes:* A higher-head, lower-flow impeller usually requires more warp or sweepback. A lower-head, higher-flow impeller requires less warp, or more nearly radial vanes. The aim of the pump designer is to arrange the vanes so as to minimize both the effect of friction on the liquid along the vanes and the radial velocity of the liquid leaving the impeller.

5. *Underfiling:* A flatter *H-Q* curve can sometimes be obtained by a method referred to as *underfiling* (Fig. 1.13). Depending on the actual pump design, increases of up to 10 percent in developed head, at or near BEP flow, can be achieved.

Pump head-flow performance can also be changed by modifying the pump casing. Some pump designs have inserts that can be added or removed depending on flow requirements. Performance can also be altered by using a process called *volute chipping* or *tongue filing*. In this process, metal is removed from the volute entrance separation wall. This affects the velocity ratio of the liquid in the casing to the liquid entering the volute.

Using underfiling and volute chipping to change performance can be risky. If the pump is already operating at or near its "ultimate" performance or design point, any alterations may adversely affect efficiency and performance.

1.6.2 Horsepower requirements

The power required to generate head in a centrifugal pump will vary with the capacity or flow rate. This can be expressed by the following relationship:

$$\text{Horsepower (liquid)} = \frac{(\text{head, ft})(\text{flow, gal/min})(\text{SG})}{3960} \qquad (1.4)$$

This is the ideal horsepower and does not take into consideration any

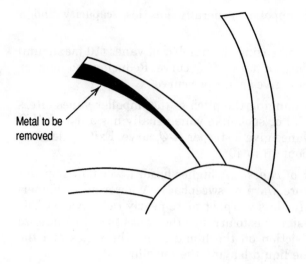

Metal to be
removed

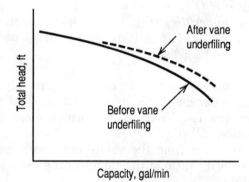

Figure 1.13   Vane underfiling to change performance.

losses within the pump. The actual or *brake horsepower* is calculated
by the following formula:

$$\text{Horsepower (brake)} = \frac{(\text{head, ft})(\text{flow, gal/min})(\text{SG})}{3960 \times \text{efficiency}} \quad (1.5)$$

This equation now includes an efficiency term that takes into account
the hydraulic shock, friction, and mechanical losses in the pump which
vary with the flow.

1.7   Affinity Laws for Centrifugal Pumps

Sometimes it is necessary to estimate the hydraulic performance of
a centrifugal pump at a speed or impeller diameter not shown on the
vendor-supplied curve. To approximate a new curve, the affinity laws
may be used.

To estimate the performance of a pump at a different speed of rotation, use the following:

1. The capacity Q is directly proportional to the ratio of the impeller speeds:

$$Q_2 = Q_1 \frac{N_2}{N_1} \qquad (1.6)$$

2. The head H is directly proportional to the square of the ratio of impeller speeds:

$$H_2 = H_1 \left(\frac{N_2}{N_1}\right)^2 \qquad (1.7)$$

3. The brake horsepower (BHP) is directly proportional to the cube of the ratio of the impeller speeds:

$$BHP_2 = BHP_1 \left(\frac{N_2}{N_1}\right)^3 \qquad (1.8)$$

For example, a steam turbine-driven pump tested at 1800 r/min gives the following results:

Capacity (gal/min)	Head (ft)	Power (BHP)	Efficiency (%)
6000	290	536	82
5500	300	508	82
5000	310	483	81
4000	335	433	78
3500	338	415	72
3000	340	368	70

To obtain the estimated performance of this pump at the reduced speed of 1200 r/min, apply the affinity laws to get these results:

Capacity (gal/min)	Head (ft)	Power (BHP)
4000	128	158
3667	132	150
3333	138	143
2667	148	128
2333	150	123
2000	151	109

The affinity laws can also be rewritten to account for changes in the impeller diameter during constant-speed operation.

1. The capacity Q is directly proportional to the ratio of the impeller diameters D:

$$Q_2 = Q_1 \frac{D_2}{D_1} \qquad (1.9)$$

2. The head is directly proportional to the square of the ratio of impeller diameters:

$$H_2 = H_1 \left(\frac{D_2}{D_1}\right)^2 \qquad (1.10)$$

3. The brake horsepower is directly proportional to the cube of the ratio of impeller diameters:

$$\mathrm{BHP}_2 = \mathrm{BHP}_1 \left(\frac{D_2}{D_1}\right)^3 \qquad (1.11)$$

Suppose the pump tested above at 1800 r/min is now driven by a constant-speed motor. If the original impeller diameter was 14.75 in, what would the expected performance be if the diameter were trimmed to 14.0 in? Applying the affinity laws yields the following results:

Capacity (gal/min)	Head (ft)	Power (BHP)
5695	261	458
5220	270	434
4746	279	412
3797	302	371
3322	304	354
2847	306	314

Test results have shown that there is a discrepancy between the calculated impeller diameter and the actual performance. For them to be equally accurate, it would be necessary for all the dimensions of the pump to be changed in proportion to the impeller change. The larger the trim, the larger the discrepancy. Therefore, impeller trims to less than 70 percent of the original (maximum) diameter should be avoided. Also, for pumps of high specific speed (2500 to 4000), impeller trims should be limited to 90 percent of the original diameter. Larger trims may result in significant efficiency drops and unstable pump performance.

1.8   Speed-Torque Relationships

Torque is the rotational force of a system and is usually expressed in terms of pounds at a 1-ft radius or, more simply, in foot-pounds (ft·lb). Torque is not dependent on rotation, but it is the moment or force tending to cause rotation. The torque T can be found from

$$T = \frac{5250 \times \text{BHP}}{n} \tag{1.12}$$

where T = torque, ft·lb
         n = speed, r/min
       BHP = brake horsepower = 33,000 ft·lb/min

To start a load and bring it up to speed, the driver must deliver more torque than required by the driven machine at any speed up to its normal running speed. The difference between the torque required by the load and the torque developed by the driver is called the *net accelerating torque*. Acceleration stops when the driven torque requirement equals the torque developed by the driver.

For pumps driven by electric motors, the motor will develop a certain amount of torque at any given speed while the pump requires another specific amount of torque (Fig. 1.14). The difference between the two is the acceleration torque which will bring the pump up to speed. Eventually, the pump and the motor will operate at the speed at which the

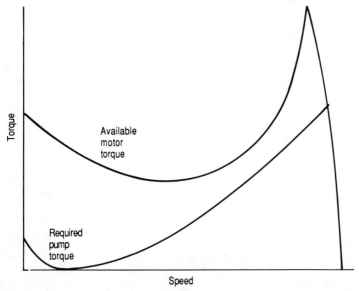

Figure 1.14   Speed-torque relationship.

speed-torque curves of the pump and motor intersect. If, however, there exists a point anywhere in the speed-torque relationship where the motor torque is less than the pump torque, then the pump and driver will slow to some reduced speed. With continued operation this may cause the motor to quickly overheat. The affinity laws, used in conjunction with a known pump performance curve, can be used to calculate a pump torque requirement as a function of speed. Once calculated, this can be compared with the proposed motor speed-torque curve to ascertain if the motor will be able to start and accelerate the pump. The speed-torque curve must be established relative to the pump flow. To minimize the starting torque requirements, it is theoretically desirable to start most centrifugal pumps against a closed discharge valve or shutoff head. However, starting a pump against a closed valve raises two potential problems: overheating of the liquid within the pump before flow is established and difficulty in opening the closed discharge valve with a large differential pressure across it. Therefore, it is recommended that the discharge valve be "cracked" or partially opened before the pump is started. Cracking open the discharge valve or starting the pump in conjunction with an opened bypass system will increase torque requirements only slightly.

2

Curves

2.1 Pump Curve versus System Curve

The basic purpose of any pump is to move a liquid from one point to another through a system of pipes, fittings, tanks, vessels, and other components. The energy required to move this liquid at a given flow rate is called the *total system head*. The total system head is commonly divided into two separate parts: suction head and total discharge head. The total head required will increase as the flow rate increases. A plot of total discharge head required versus flow rate generates a system curve, against which all pumps are required to operate. We do not go into the details of how a system is analyzed and a system curve generated. There are several handbooks and computer programs that do an excellent job of dealing with this. However, a clear understanding of the interaction between the system curve and the centrifugal pump performance curve is necessary to obtain reliable, efficient operation and to assist in field troubleshooting.

The total head versus flow rate curve of a system is made up of three basic components:

Static head

Pressure head

Friction head

Figure 2.1 shows how these components combine to create a system curve. As can be seen, the static head and pressure head components are constant at all flow rates. However, the frictional head increases as the flow rate increases. In addition, in most systems, there is a control valve which is adjusted to add or remove frictional head and thus controls the pump flow rate. Generally, the first step in generating a

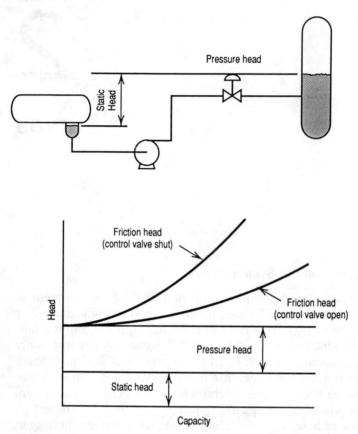

Figure 2.1 System curve.

system curve is to draw a one-line system sketch from which the *net positive suction head available* (NPSH$_a$) and discharge head requirements can be calculated.

Once a system curve has been established, a pump performance curve can be selected to match the system curve (Fig. 2.2). Remember, the pump will always operate at the intersection of the pump curve and the system curve, and the system curve controls the flow rate through the pump, not the other way around.

2.2 Creating a System Sketch

System sketches are a means of graphically describing new or existing piping arrangements (Fig. 2.3). However, they are also valuable for many other purposes, such as selecting and sizing piping and control valves, determining the vessel skirt height, and assisting in setting

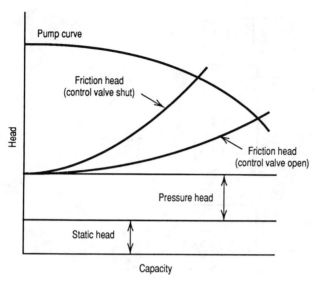

Figure 2.2 Pump curve and system curve interaction.

thermosiphon reboiler elevations. It is very important that the information on the system sketch be presented in a consistent and accurate manner, using a format that is easily understood now or in the future. The following guidelines are presented to assist in the completion of a system sketch. When actual pressure drops, elevations, and other field data are known, those values should normally be used. Figures 2.13 to 2.20, which can be found at the end of this chapter, provide examples of several completed system sketches for various frequently encountered process systems.

2.2.1 General information needed

Figure 2.4 is a typical data form that can be used to develop a system sketch and to calculate the suction and discharge heads. Here are some guidelines and rules of thumb:

1. Normally, the source for the information required to complete a system sketch is available either from an initial process design report, existing unit data books or from actual field measurements.

2. Each individual pumping or major process system should be analyzed separately by completing a system sketch. Additional sketches showing alternate operating conditions, such as significant turndowns, seasonal changes, and common-spare operation, should be completed. On each sketch show the expected percentage of operating time for all alternatives.

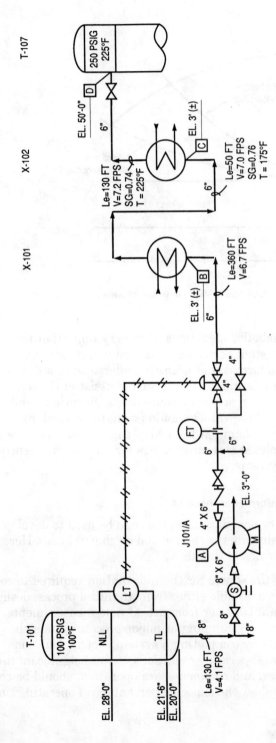

Figure 2.3 Typical process system sketch.

SUCTION PIPING CALCS.

SUCTION PIPING CALCS.	GPM	SUCT.P	NPSHg	DISCHARGE PIPING CALCS.	BRANCH
RATED FLOW @ P.T.	GPM			NORMAL FLOW @ T.	(GPM)
SPEC. GRAV. @ P.T.					
P: PRESS. ABOVE LIQUID	PSIG			TERMINAL PRESSURE	(PSIG)
Pa: ATMOS. PRESSURE	PSIA			FRICTION LOSSES	
Pv: VAPOR PRESSURE	PSIA			PIPING LOSSES (+)	(PSI)
P + Pa - Pv =		PSI		HEATER (+)	(PSI)
PIPING FRICTION LOSSES		PSI		HEAT EXCHANGER (+)	(PSI)
OTHER FRICTION LOSSES		PSI			(PSI)
L: TOTAL FRIC. LOSSES (-)		PSI			(PSI)
S: STATIC HD. FROM GRD.		FT			(PSI)
B: PUMP C/L TO GRADE		FT		TOTAL FRICTION LOSS	(PSI)
NET STATIC HEAD (+/-)		FT		STATIC HEAD (+/-)	(PSI)
NPSH AVAIL. W/O MARGIN		FT		CONTROL VALVE (+)	(PSI)
M: MARGIN (-)		FT		ORIFICE PLATE (+)	(PSI)
NPSH AVAILABLE		FT			
PUMP SUCTION PRESSURE		PSIG		DISCHARGE PRESSURE	(PSIG)

LOCATION _____ UNIT _____
ITEM NO. _____
FLUID _____ SERVICE _____
PUMPING TEMP. _____ °F ATMOS. PRESS. _____ PSIA
SPECIFIC GRAVITY @ 60°F/P.T.: _____ / _____
VAPOR PRESS. @ P.T. _____ PSIA
VISCOSITY @ P.T. _____ SSU _____ CP
CORROSION/EROSION CAUSED BY _____
FLOW @ P.T.: MINIMUM _____ GPM
NORMAL _____ GPM
RATED _____ GPM

PUMP PARAMETERS
PRESS.: DISCH _____ PSIG SUCT. _____ PSIG
DIFF. PRESS. _____ PSI _____ FT
NPSH AVAIL. _____ FT
MAXIMUM ALLOWABLE SYSTEM PRESS. _____ PSIG
AS LIMITED BY _____
FOR ALTERNATE OPERATING CONDITIONS, SEE PIPING SYSTEM SKETCH NUMBER(S) _____
FLOW SHEET REFERENCE _____

REVISION	DATE	BY

PIPING SYSTEM SKETCH SHEET
(FOR PUMPS)

PROJECT NO. _____ APPN. NO. _____
PREPARED BY _____
APPROVED BY _____
DATE _____
TITLE _____
SKETCH NO. _____ REV

Figure 2.4 Typical system sketch data form.

3. Use only one system sketch for a combination of a main and a spare pump. Use separate sketches where a pump is a common spare for two different services.

4. The pumping temperature is the fluid temperature at normal operating conditions. If unusual conditions will affect the fluid temperature and/or viscosity, such as during start-up, a separate system sketch for those conditions should be completed.

5. The specific gravity of a fluid and thus the horsepower required to pump it vary with temperature. Therefore, it is important to know the range of the specific gravity from normal pumping temperature to ambient temperature (for pump selections this is normally assumed to be 60°F). However, if there will be any operating conditions that could significantly affect the specific gravity beyond these two temperatures (e.g., taking cold suction from a storage tank, start-up, etc.), additional system sketches should be completed. Specific-gravity values are normally provided in the process design report; or for most refinery and petrochemical fluids, reasonable values can be obtained from one of several industry-accepted handbooks.

6. The vapor pressure of the pumped fluid is the pressure of the gas above the liquid, in a closed vessel, at normal pumping temperature. Normally, unless it is specifically known or stated otherwise in the process design report, assume that the gas is in equilibrium with the liquid in that vessel.

7. Viscosity is a measure of that property of a fluid which offers resistance to flow because of the existence of internal friction in the fluid. The viscosity of a fluid affects not only the pressure drop through a piping system but also the horsepower required to pump it. Performance curves for most pumps are based on pumping water. Therefore, when the fluid being pumped is very viscous, the mechanical equipment application engineer must apply correction factors to these curves to determine the actual expected field performance and the required driver horsepower. As with specific gravity, if the viscosity will vary significantly from start-up to normal pumping conditions, this should be noted and a separate system sketch prepared for the alternate case.

8. The applications engineer must know if the fluid to be pumped is corrosive or contains particles and solids that could cause erosion of internal components of the pump. Based on this information, the equipment engineer will select the proper metallurgy for the pump and recommend specific impeller designs, tip speeds, and sealing arrangements.

9. System designers like to talk about flow rates in terms of the *minimum flow, normal flow,* and *rated flow.* The minimum flow is the lowest flow rate that the pump will see under *any operating* or start-up

condition. This flow rate is a process parameter, *not* a pump design criterion. The minimum flow required by a centrifugal pump (Chap. 3) is limited by the pump design. For a conservative approach, assume it is limited to 45 percent of the BEP flow rate. To prevent possible damage to the pump, a minimum flow recirculation line is often used which bypasses a portion of the flow from the discharge of the pump back to the suction vessel. Although the final decision concerning the need for a minimum flow recirculation bypass cannot be made before the pump has been selected, the potential need for one can be identified. Normally the application engineer will select a pump such that the BEP flow rate lies between the normal and rated flow. Therefore, if the minimum *process* flow is less than 45 percent of the normal flow rate, a minimum flow recirculation line is normally required.

Rated flow refers to the flow rate which the pumps and other equipment in the system should be capable of handling in a like-new condition. Unless specifically stated in the process design report, assume that the rated flow is 110 percent of the normal flow rate. This service factor is added to provide for short-term variations in operating conditions and long-term deterioration of the pump efficiency due to wear. To prevent the possibility of grossly oversizing the pump, no other additional service factor should be added.

10. The discharge pressure and the suction pressure of the pump and the $NPSH_a$ can be calculated by completing a system sketch. The differential pressure is the difference between the discharge and suction pressures. The differential head is the amount of energy (in feet of fluid) the pump must impart to the fluid. It can be calculated as follows:

$$\text{Differential head} = \frac{2.31 \times \text{differential pressure}}{\text{SG}} \qquad (2.1)$$

11. Each piping system is made up of several pieces of equipment, vessels, heat exchangers, piping, etc., each of which has a maximum pressure rating. It is very important that the shutoff pressure of the selected pump not exceed the allowable pressure rating for the most limiting downstream component (most centrifugal pumps develop the maximum differential head or discharge pressure at or near shutoff or zero flow point). For new systems, the design pressure is normally determined by the shutoff head of the pump. When existing systems are retrofitted, use of larger pumps quite often results in situations where the shutoff head exceeds the maximum allowable pressure of some downstream component. In these cases, if an alternate pump having a lower shutoff head and still meeting the flow requirements cannot be found, the downstream equipment must be replaced, rerated, or protected by relief valves. It is very important that the weak link in

a system be determined as early as possible in the project design. Failure to do so may result in significant cost increases, delays for the project, or, if not corrected, a potential safety hazard in the field. Data on the maximum allowable pressure for each component can usually be found in the unit data books, on equipment nameplates, or from the manufacturer.

The actual shutoff pressure will not be known until the pump is selected. Therefore, to determine if there is a potential problem, the shutoff pressure should be estimated. The differential pressure developed by most low-pressure (less than 250 lb/in^2 differential pressure) centrifugal pumps rises approximately 10 to 15 percent from that at the normal flow rate to shutoff. For higher-pressure differential pumps, the rise to shutoff is typically only 5 to 10 percent.

2.2.2 Creating a system sketch

1. Sketch the piping system schematically (Fig. 2.3). Each sketch will normally begin and end at a vessel or tank having a controlled pressure. Ordinarily, this would include as a minimum a supply vessel, a pump, suction and discharge piping, and a terminal vessel, tank, or drum. The sketch should include all flow-measuring devices (orifices, rotameters, etc.), control valves, block and check valves, strainers, filters, exchangers (with indication of shell or tube side flow), furnaces, in-line mixers, spray nozzles, etc. Elbows, vents, and drains need not be shown.

2. Show on the system sketch the process fluid temperature and pressures. A correct system sketch requires the use of consistent pressure notation throughout. It is surprising how frequently a pump suction pressure is given in pounds per square inch absolute and the discharge in pounds per square inch gage. These are then directly combined to give an erroneous differential pressure. Normally, only the atmospheric pressure and vapor pressures are given in pounds per square inch absolute. All other pressures should be in pounds per square inch gage.

3. All pertinent elevations should be shown on the system sketch, i.e., the height above a fixed reference level or datum plane (usually called the *grade level*). These heights or elevations should include vessel tangent lines, liquid levels (high, normal, low), bottom of vessel or drum, feed or reflux nozzle elevations, pump centerlines, etc. Also show the elevations of the start and endpoint for all branch lines. A good source of data for making these estimates is the unit data books for other similar but existing units. When the actual pump centerline is not known, assume that it is 3 ft above grade.

4. Identify all piping branches or sidestreams. A different branch should be designated at each piece of equipment, at changes in the flow patterns (from series to parallel flow), and when the process fluid properties change (such as through a furnace or heat exchanger). Show all specific gravities or specific weights.

2.2.3 Determining the pressure drop due to piping

1. Pressure drops occur throughout the system because of resistance or friction within the piping, fittings, and other components; the throttling action across control valves; and the differences in height of the pumped fluid. All these changes in fluid pressure must be accounted for in order to properly size the pump and its driver to meet the design conditions. The pressure changes should be calculated for each branch in the suction and discharge piping, and the values shown on the system sketch.

2. The pressure drop due to resistance or friction depends on many things, including the size and type of pipe and fittings being used, condition (roughness) of the pipe's interior surface, length of pipe, and characteristics of the fluid, i.e., viscosity and flow rate.

3. By knowing the flow rate, fluid characteristics, and pipe diameter used, the pressure drop per 100 ft can be calculated. The total pressure drop can be calculated by using equivalent lengths of the system. The equivalent length is the length of straight pipe plus the equivalent length of the fittings and other mechanical components. The equivalent length for various fittings can be found in either the Cameron[1] or Crane[2] handbook while the equivalent lengths for other mechanical equipment can usually be found in the manufacturer's data book.

4. For an existing system, a field survey will provide the exact number and size of pipe and fittings. For new systems, if completed piping drawings or models are available, they should be used to obtain the actual length of pipe, number of valves, fittings, etc., to determine the equivalent length. However, usually the piping drawings are not available when the system sketches are being prepared. It is then necessary to arrive at an estimated equivalent length by scaling the plot plan and elevation drawings (if available) and estimating the number of valves, tees, ells, reducers, etc. The assumed routing should account for typical process plant layouts, i.e., from equipment into a pipe rack and out to another piece of equipment, unless piping preliminary layouts indicate otherwise. Piping flexibility and the need for expansion loops should also be considered.

For most unit process piping, the discharge piping friction loss, compared to the pressure drop across the control valve, is only a minor part of the total pump head required. Therefore, a high degree of accuracy is not required in the friction loss calculations. If the lines are more than a few hundred feet long or if mixed-phase or nonnewtonian flow is predicted, more sophisticated calculation techniques are required.

2.2.4 Pressure drop due to other system components

1. Determine the pressure drop across major process equipment such as fired heaters, reboilers, etc. For existing equipment, use known field data. For new or revamped equipment, pressure drop data can usually be obtained from the equipment manufacturer or vendor catalogs. If not, they can be estimated based on similar existing equipment or process designs.

2. The pressure drop across a heat exchanger can be obtained from existing field data, manufacturer's design data, or calculated values provided in the process design report. If no other information is available, an estimated value of 10 lb/in^2 is normally adequate for the tube side and 5 lb/in^2 for the shell side (per shell).

3. For rotameters, positive displacement meters, and orifice-type flow meters in liquid service 2 lb/in^2 is the normal estimated pressure drop.

2.2.5 Control valve pressure drop

1. The pressure drop across a control valve is a very important aspect of analyzing a system sketch. The point of sizing control valves is to specify a minimum pressure drop for energy conservation, but sufficient for adequate control of the process.

2. The pressure drop across a control valve (liquid service) is based upon the valve flow coefficient index C_v as follows:

$$PD = SG\left(\frac{Q}{C_v}\right)^2 \tag{2.2}$$

where PD = pressure drop, lb/in^2
    SG = specific gravity
    Q = flow, gal/min
    C_v = valve flow coefficient

2.2.6 Static head pressure changes

1. Static head is the change in system pressure as a result of a change in the elevation or height of the fluid being pumped. To calculate the static head, use the following:

$$H_s = \frac{(\Delta h)(SG)}{2.31} \qquad (2.3)$$

where H_s = static head, lb/in^2
Δh = change in fluid height, ft
SG = specific gravity

2. Note that the specific gravity can change significantly with changes in fluid temperature. The specific gravity given in the process design report is usually at normal pumping temperature. For systems having heat exchangers, furnaces, reboilers, etc., the gravity will change across each piece of equipment. Therefore, to determine the overall change in pressure due to static head, calculate the static head for each segment (of constant temperature) and algebraically add the results to obtain the overall static head.

3. Static head may either be added to or subtracted from the system pressure. For suction piping where the pump is taking suction from an elevated vessel (above the suction centerline), static head is added to the source pressure. However, if the pump is taking suction from a pit or well (suction lift), the static head is subtracted from the source pressure. In discharge piping, when the pump discharges into an elevated vessel, the static head is added to the terminal pressure for that branch line. At times, there may be sidestreams from some other source that combine with the main discharge flow downstream of the pump. Calculate static head for this sidestream from the point where it joins with the main flow back to its origin.

2.2.7 Guidelines for sizing suction piping

1. Because of $NPSH_a$ considerations, the suction piping requires a more accurate evaluation than the discharge piping. Sizing should be based on the *rated* flow.

2. Suction lines should not normally be smaller than the pump suction nozzle.

3. To prevent cavitation in the pump, suction line velocities should not exceed 10 ft/s unless there is a substantial length (at least six pipe diameters) of straight pipe immediately upstream of the pump. Consider 5 to 6 ft/s maximum for new systems.

4. The pressure drop across permanent suction strainers must be considered.

2.2.8 Guidelines for sizing discharge piping

1. Sizing should be based on *normal* flow.

2. In general, the most economic line size is obtained by balancing the initial installation cost against the operating cost for various line

sizes. In most cases, the use of smaller line sizes will increase the operating costs because of the increased pump horsepower requirement needed to overcome the associated higher system pressure drop.

3. Lower flow velocities should be considered for reduced power consumption and when entrained particulate material may cause erosion. For slurry systems, the fluid velocity should be kept between 5 ft/s to prevent settling and 9 ft/s to prevent erosion.

2.3 Calculating NPSH$_a$

2.3.1 General

Accurate determination of the NPSH$_a$ is necessary to ensure good operation in the field. The formula for calculating the NPSH$_a$ varies depending on the configuration of the suction system. However, the general equation is as follows:

$$\text{NPSH}_a = \frac{(P + P_a - P_v)(2.31)}{\text{SG}} + \left(\pm S - B - L + \frac{V^2}{2g}\right) \quad (2.4)$$

where NPSH$_a$ = net positive suction head available, ft
P = pressure above liquid, lb/in^2 gage
P_a = atmospheric pressure, lb/in^2 atmospheric
P_v = vapor pressure of liquid at pumping temperature, lb/in^2 atmospheric[*]
SG = specific gravity at pumping temperature
S = static height of liquid above (+) or below (−) grade[†]
B = distance of pump centerline (suction nozzle centerline for vertical pumps) above grade, ft
L = suction system friction losses, ft (of liquid)[‡]
V = average liquid velocity at pump suction nozzle, ft/s[§]

---

[*]When the liquid is blanked or pressurized by a gas that is different in composition from the liquid, the effect on the true vapor pressure due to absorption must be determined and used in the calculation. If specific information is not available, treat the calculation as if the vessel contained a boiling liquid.

[†]The static height of liquid above grade used in the calculation varies with the type of system and application; see subsequent examples for specific information.

[‡]Suction system losses should include piping pressure drop and the pressure drop of partially fouled suction screens and strainers.

[§]The quantity $V^2/(2g)$ is frequently deleted from the NPSH$_a$ calculation because it normally has an insignificant effect.

2.3.2 Margin

To account for unforeseen differences between the actual NPSH_r and the vendor catalog values and the actual NPSH_a and that calculated, frequently a margin is utilized. For most hydrocarbon streams, the margin used is normally 0 to 5 ft. However, for some special services, e.g., boiler water feed and vacuum tower bottoms, a margin of up to 10 ft is commonly used.

The addition of a margin M and the deletion of the $V^2/(2g)$ term result in

$$\text{NPSH}_a = \frac{(P + P_a - P_v)(2.31)}{SG} + (\pm S - B - L) - M \qquad (2.5)$$

2.3.3 Pumps taking suction from above-grade vessels and drums

If the normal operating liquid level is 5 ft or more above the bottom of the vessel (10 ft for special services), the available NPSH should be calculated from the bottom of the vessel (Fig. 2.5).

$$\text{NPSH}_a = \frac{(P + P_a - P_v)(2.31)}{SG} + (S - M - B - L) \qquad (2.6)$$

If the normal operating liquid level is less than 5 ft (10 ft for special services) above the bottom of the vessel, the available NPSH should be calculated from an elevation 5 ft (10 ft for special services) *below* the

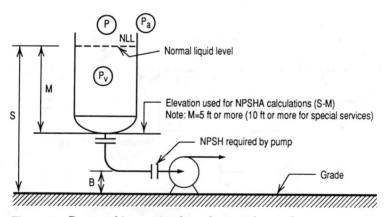

Figure 2.5 Pumps taking suction from above grade vessels.

$$\text{NPSH}_a = \left[\frac{(P + P_a - P_v) \times 2.31}{SG}\right] + (S - M - B - L)$$

normal operating liquid (Fig. 2.6). Formula (2.6) applies. However, for boiling liquids, the value of $(P + P_a - P_v) \times 2.31 / SG$ is zero and

$$\text{NPSH}_a = S - M - B - L \qquad (2.7)$$

2.3.4   Pumps with a suction lift

Pumps with a suction lift should have an NPSH margin of at least 5 ft based on the normal liquid level. Where the variation between the normal liquid level and the minimum liquid level is 5 ft or more, the available NPSH should be based on the minimum liquid level; where the variation is less than 5 ft, the available NPSH should be based on a level 5 ft below the normal liquid level (Fig. 2.7).

$$\text{NPSH}_a = \frac{(P_a - P_v)(2.31)}{SG} - (S + M + B + L) \qquad (2.8)$$

2.3.5   Suction from atmospheric storage tanks and cooling tower basins

The available NPSH for pumps taking suction from atmospheric storage tanks usually is calculated from the top of the suction nozzle in the tank (Fig. 2.8). Since the tank level normally is 5 ft or more above the tank suction nozzle, and the pumping rate usually is reduced if the tank level approaches the suction nozzle level, no margin is normally included in the NPSH_a calculation.

$$\text{NPSH}_a = \frac{(P_a - P_v)(2.31)}{SG} + (S - B - L) \qquad (2.9)$$

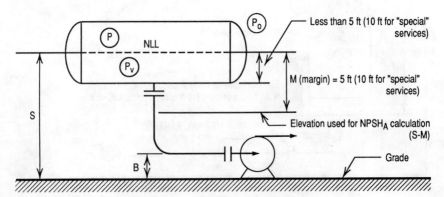

Figure 2.6   Pumps taking suction from above grade drums.

$$\text{NPSH}_a = \left[\frac{(P + P_a - P_v) \times 2.31}{SG} \right] + (S - M - B - L)$$

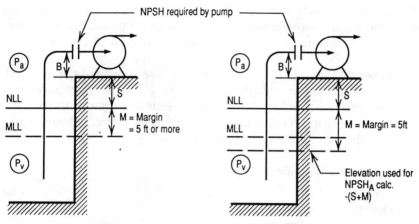

Figure 2.7 Pumps with a suction lift.

$$\text{NPSH}_a = \left[\frac{(P_a - P_v) \times 2.31}{SG} \right] - (S + M + B + L)$$

Where the pumping system does not include a specific flow controller, such as most transfer services (pumping from one tank to another) and some loading services, an analysis is required to determine the maximum system flow for determining NPSH_a, for specifying the rated flow of the pump, and for establishing the NPSH required by the pump at this maximum flow rate. The analysis should include

- Pumping from a full tank (maximum NPSH_a)
- Pumping from an empty tank (minimum NPSH_a)
- Pumping to an empty tank (maximum flow rate)

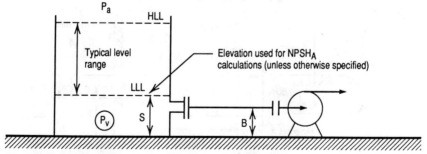

Figure 2.8 Pumps taking suction from atmospheric storage tanks.

$$\text{NPSH}_a = \left[\frac{(P_a - P_v) \times 2.31}{SG} \right] + (S - B - L)$$

Where the composition, temperature, and/or vapor pressure of the fluid in the tank varies, the maximum realistic vapor pressure (and viscosity when applicable) must be determined and used in the NPSH calculations.

For cooling tower basins (Fig. 2.9), the $NPSH_a$ formula is Eq. (2.9). However (Chap. 3), in addition to available NPSH submergence must be considered to avoid vortex formations in the area of the suction nozzle. The vapor pressure used in calculating $NPSH_a$ should correspond to the maximum basin temperature.

2.3.6 Vertical sump pumps

Submergence (Chap. 3) of the suction end bell must be taken into consideration in addition to the available NPSH for vertical sump pumps. The submergence requirement usually establishes the sump depth and minimum liquid level at the maximum flow rate. Submergence requirements are usually available from the pump vendor and are generally based on a recommended sump design. A typical sump pump installation is shown in Fig. 2.10. For sump pumps, the formula for $NPSH_a$ is

$$NPSH_a = \frac{(P_a - P_v)(2.31)}{SG} + (Q - B) \qquad (2.10)$$

2.4 Normal and Rated Capacity versus Efficiency

In the past, it was rather common to select both normal and rated capacity to the left of the pumps' capacity at peak efficiency. The two major factors that were cited as justification for such selections were

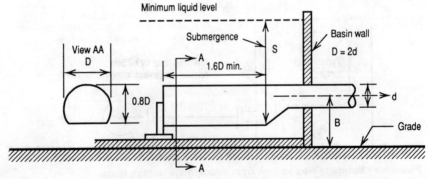

Figure 2.9 Pumps taking suction from grade mounted cooling tower basin.

$$NPSH_a = \left[\frac{(P_a - P_v) \times 2.31}{SG} \right] + (S - B - L)$$

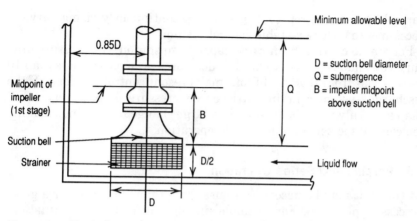

Figure 2.10   Vertical sump pumps.

$$\text{NPSH}_a = \left[\frac{(P_a - P_v) \times 2.31}{\text{SG}} \right] + (Q - B)$$

the future expendability of pump capacity and lower NPSH requirements. However, such selections have major disadvantages, including higher operating costs, greater shaft deflections due to radial thrust, and higher probability of hydraulic system pulses due to eye or tip recirculation within the impeller. Power costs have increased such that any selection criteria based on operating costs would require that rated capacity equal the best efficiency flow. However, because pumps are normally only available in fixed frame sizes, this is seldom possible.

Today, many users have adopted the philosophy that the normal flow capacity should fall to the left of the peak efficiency flow and the rated flow to the right. The merits of this selection criterion were recognized by API and is now a preferential requirement of API Standard 610, *Centrifugal Pumps for General Refinery Service.* Due to the wide variation in required pump capacities and the limited number of pump sizes and impeller patterns, not all applications will meet this selection criterion. However, the general aim is to minimize power consumption by selecting pumps with the projected normal capacity as close as practical to maximum efficiency.

Pump head-capacity curves vary in shape from relatively flat to rather steep and may even be humped. For some applications, with high discharge system friction drop, the humped curves are completely acceptable; but on other systems having little or no friction drop, the humped curve is unacceptable. API 610 requires head-capacity curves which continuously rise to shutoff for all pumps specified for parallel operation, and they are *preferred* for all other services. However, many applications of single operating pumps also require a continuously rising curve. For example, when the capacity will be controlled by a variable pressure drop control valve, a continuously rising curve is necessary to ensure a stable control system.

For stable control, from rated capacity to shutoff, the pressure drop must continuously increase as flow decreases. Flat curves (less than 10 percent rise from BEP to shutoff) may present control problems. If the discharge system contains little or no pressure drop, a very small change in valve position results in a large change in flow, thus making the control too sensitive for stable operation.

2.5   Parallel and Series Operation

At times, it may be necessary to use more than one pump for a given process application. For example, in high-head, low-flow situations such as many pipeline applications, it may be desirable, due to cost, to use two pumps in series than to develop the total system head requirements in one pump. For high-flow, low-head applications, such as commonly found in cooling tower circulating water pump service, it is more desirable to use several smaller-capacity pumps in parallel to produce the required flow than to use one large-capacity pump. This is generally true when flow requirements can vary greatly or when flow reliability demands multiple parallel pumps. Figure 2.11 shows the combined head-capacity curve of two pumps in series. For two pumps in series, the developed head at any given capacity is added to produce the head-capacity curve for the two pumps. Figure 2.11 also shows the combined head-capacity curve for two pumps in parallel. In this case, the capacity at any given developed head is added to yield the head-capacity for the two pumps in parallel. There are several potential problem areas associated with operating two pumps in series.

1. There is the possibility of overpressurizing the downstream pump. Remember that the second pump in a series operation has a suction pressure equal to the discharge pressure of the first pump and that the pressure ratings of the downstream pump and its sealing devices must be capable of operating safely at the *combined* shutoff pressure of the two pumps.

2. If for some reason the downstream pump is stopped, then either the upstream or booster pump must also be secured or, to prevent damage, provisions must be made to maintain a minimum flow through the pump.

3. To prevent starving the downstream pump, make sure that the booster pump's performance curve adequately covers both the system and downstream pump performance curves for all potential flow conditions and NPSH$_r$ requirements.

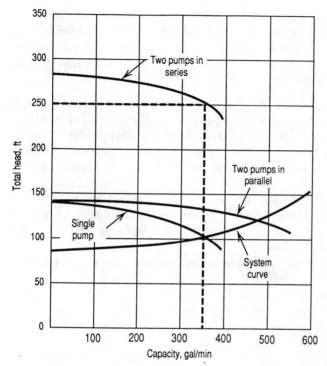

Figure 2.11 Parallel and series operation.

In parallel operation, the major problem is load sharing.

1. When the head-capacity curve of each pump in a parallel config-uration is identical, the problem of load sharing is minimized. How-ever, even with pumps with supposedly identical head-capacity curves, the dangers involved in load sharing are always present and low-flow protection is extremely important.

2. Frequently pumps are operated in parallel to ensure reliability of flow. To improve this reliability, one of the pumps may be steam-turbine-driven while the other is driven by a "constant-speed" motor. However, two load-sharing problems can arise from this situation due to the speed differential between the two pumps. First, the motor will operate at a slip frequency that will vary depending on the load; this may or may not equal the turbine speed. Second, depending on the NEMA class governor used on the steam turbine, its rotational speed may vary as much as ±10 percent with load changes. In either case, the two pumps will fight each other, backing one up on its performance curve.

3. There is no such thing as an "identical pump." Each pump has its own performance curve which is a result of all the minor casting and

machining discrepancies built into it as well as the changes made in it during operation due to corrosion and wear. It is not uncommon for two pumps to operate satisfactorily in parallel for many years and then to experience problems after one of them has been overhauled; a new impeller does not perform exactly as the old one did.

The dangers of load sharing can be demonstrated by considering parallel operation of two nonidentical pumps. Figure 2.12 shows this. Both pumps will operate at the intersection of the combined pump curve and the total system curve. When the system control valve is throttled to reduce total flow, the head generated by pump B quickly equals the head generated by pump A at zero flow, while pump B is operating near half-capacity. This means that pump A has been forced off the line, and if appropriate minimum flow recirculation protection is not present, the pump may be destroyed. Pump A and pump B do not share the load well together.

Special considerations need to be addressed when series and parallel pumps are used in pipeline applications. The ideal pump station would have only one pump operating at its BEP. However, pressure, capacity, or horsepower limitations frequently dictate a need for multiple pumps in series or in parallel. Series operation, when each pump delivers full throughput and generates partial pressure, has considerably more flexibility than the comparable parallel installation. If necessary, pumps can have their impellers trimmed to match the system requirements with a minimum of throttling. Parallel pumps are less flexible since each pump must generate the maximum pressure while deliver-

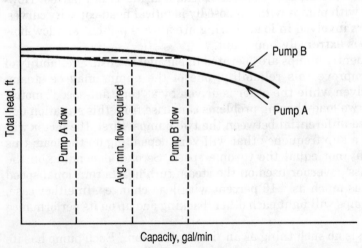

Figure 2.12 Parallel operation with mismatched performance curves.

ing only partial throughput. Thus, only the capacity of the impeller can be varied.

It is reasonable to state that when friction losses make up most of the total system head requirements, series pumps are the logical choice. In the event one of the pumps fails, partial pipeline throughput can normally be maintained by using the remaining pump(s). However, when the pressure head makes up the greater proportion of the total head required, parallel pumps become the more logical choice; again, if one pump fails, the pipeline will be able to maintain partial pressure.

Suction piping design   Improperly sized and tortuous flow paths of suction piping can be disastrous. Sometimes piping fittings such as block valves, check valves, and strainers are stacked vertically to accommodate pipe rack space limitations. This practice can create noise and vibration in the pump. Piping tees, valves, strainers, and elbows cause flow disturbances that can be especially harmful to double suction impellers. These good practices can be used to reduce these problems on the suction piping:

- Valve stems and tee branches should be installed perpendicular to the pump shaft, not parallel to it.
- Piping should have at least five pipe diameters of straight run before the suction flange.

Discharge piping design   Fluid flow disturbances in valves and piping on the discharge side of a pump can create vibrational waves which travel freely in the piping, transfer to pipe hangers, and appear at the pump. Loose-fitting valve gates or plugs and excessive throttling by use of either the discharge valve or a control valve frequently cause problems. Dynamic pipe reaction frequently imposes high vibration forces on pump casings. Heavy forces imposed by pipe may sufficiently distort casings and bearing housings to create vibration. Provisions must be made to independently support piping and provide for thermal expansion of both the pump and its piping. This is especially important in hot service pumps. New pipe design on pumps operating above about 400°F should be carefully reviewed or audited by the owner or other independent party for correct input and desired results prior to construction.

Piping vibration limits   How much vibration can be permitted in the piping system? One rule of thumb, which works very well, states that the permissible unfiltered velocity readings taken on the piping at the midspan of its supports can be 3 times the permissible readings taken on the bearing caps. Experience indicates that bearing cap readings from 0.5 to 0.6 in/s are cause for concern for a pump and 1.0 in/s is the emergency shutdown level. These pump vibration guidelines then give 1.5 to 3.0 in/s as the limit for piping.

References

1. *Cameron Hydraulic Data,* Ingersoll-Rand Company, Woodcliff Lake, N.J., 1979.
2. *Flow of Fluids Through Valves, Fittings, and Pipe,* Technical Paper No. 410, Engineering Division of Crane, Crane Co., Chicago, Ill., 1978.

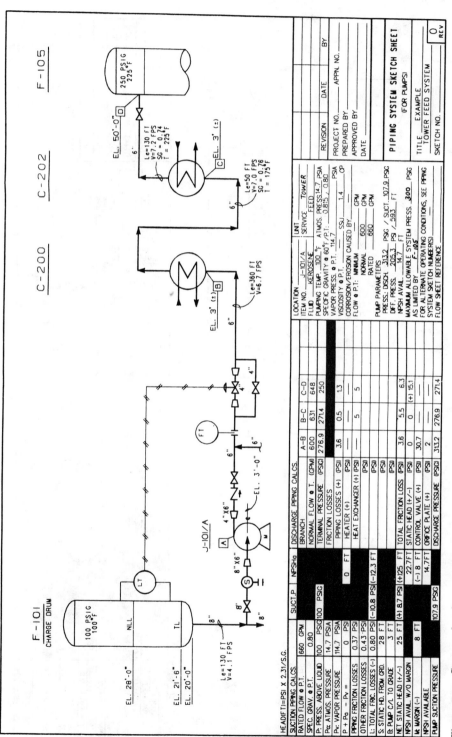

Figure 2.13 System sketch; typical feed system.

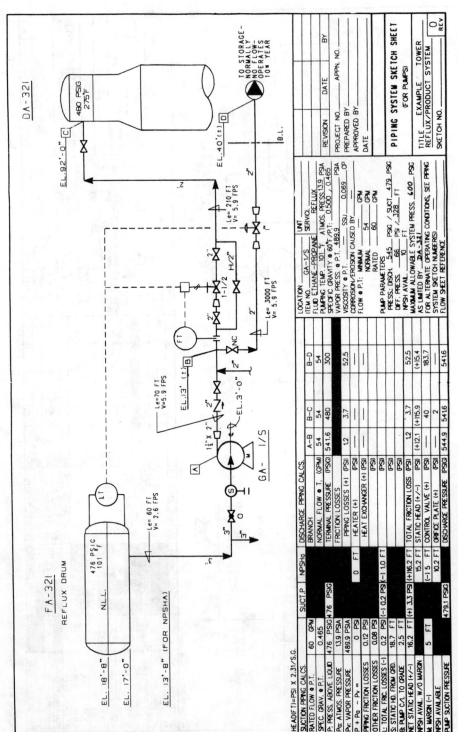

Figure 2.14 System sketch; typical reflux system.

48

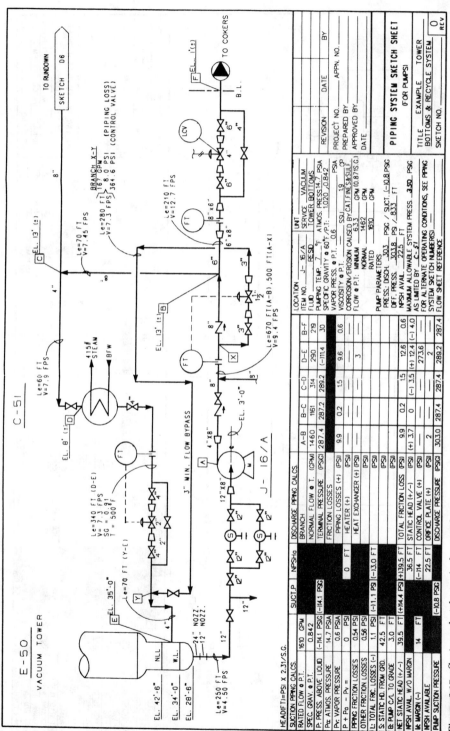

Figure 2.15 System sketch; typical tower bottoms system.

49

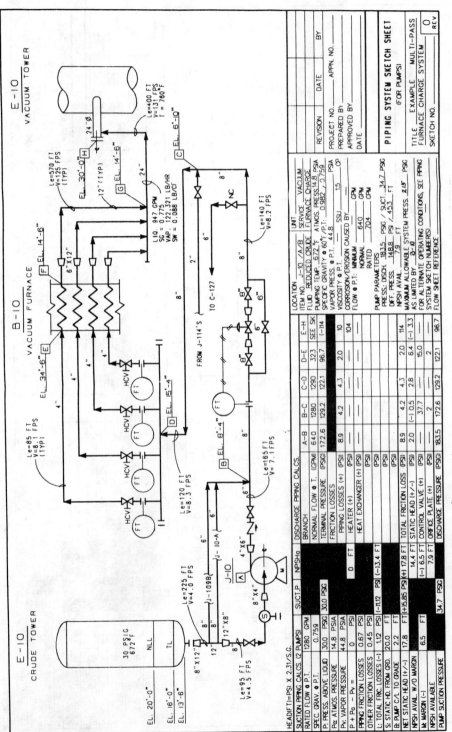

Figure 2.16 System sketch; typical furnace charge system.

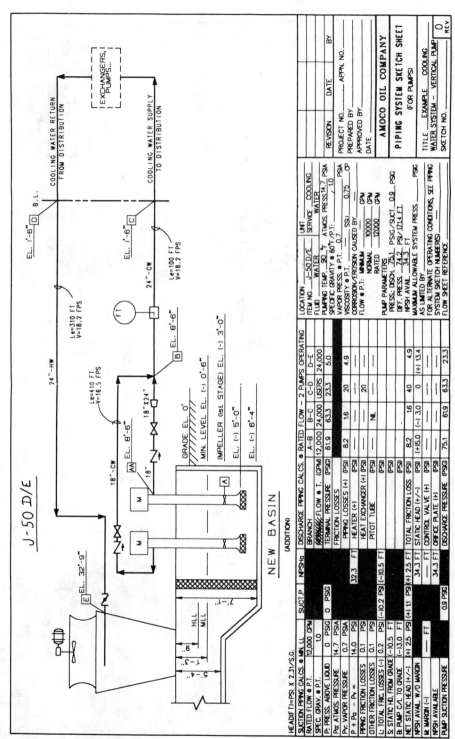

Figure 2.17 System sketch; typical vertical pump system.

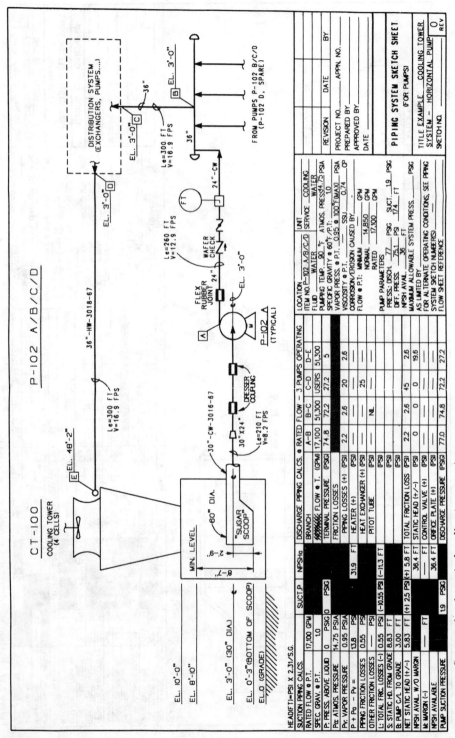

Figure 2.18 System sketch; typical cooling tower system.

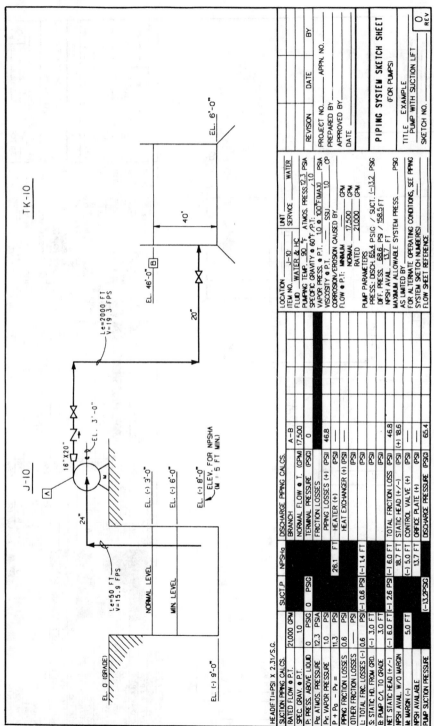

Figure 2.19 System sketch; typical sump pump system.

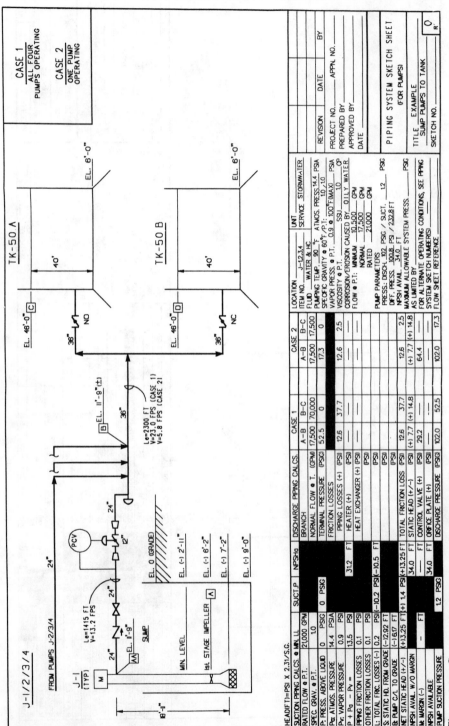

Figure 2.20 System sketch; typical suction lift system.

Suction Performance

In the preceding chapters we reviewed the pump head-capacity curve, its interaction with the system curve, and specific speed. So far, all the discussion has related to the pump discharge performance. However, equally important is the pump suction performance.

3.1 NPSH, a Major Factor in Pump Operation

The term *net positive suction head* (NPSH) is widely used and just as widely misunderstood. Most pump handbooks are loaded with NPSH calculations that treat this suction condition as a static situation when, in fact, it is dynamic. Then what is NPSH?

Pumps are wonderful machines that do an excellent job of moving a fluid and raising its pressure as long as that fluid is a liquid and remains a liquid while in the pump. Any gas or vapor formed in or passing through a pump will reduce the hydraulic performance and potentially may be physically damaging to the pump itself. But how do gases and vapors come to be in the liquid being pumped, and more importantly, how can we keep them out?

3.1.1 NPSH and cavitation—The classical approach

A pure liquid, such as water, held at a constant temperature will vaporize or boil if its pressure is sufficiently reduced. The pressure at which the liquid vaporizes is called its *vapor pressure*. For example, water at 100°F will boil or vaporize if exposed to a vacuum of 28 inHg. When vaporization occurs within a pump, it is called *cavitation*. Keep in mind that very few process applications call for a pump to handle a pure liquid like water. Most services handle a mixture of various components such as crude oil, blended gasoline, or even paint, and as such they will have a range of vapor pressures or boiling points for each component.

The pressure of the liquid in a centrifugal pump drops as it flows from the suction flange, through the suction nozzle, and into the impeller. The amount of pressure drop is a function of many factors, including the design geometry, rotational speed, frictional and hydraulic shock losses, flow rate, etc. (Fig. 3.1). If, however, that pressure, at any point within the pump, falls below the vapor pressure of the liquid being pumped, then vaporization or cavitation will occur.

A curve that shows the minimum pressure (commonly expressed in terms of head measured at the suction flange) required to prevent vaporization anywhere in the pump versus the flow rate through the pump can be generated for any pump-impeller combination. That minimum required head, for any given flow rate, is called the *net positive suction head required* (NPSH$_r$).

In a pumping system, the difference between the actual pressure or head of the liquid available (measured at the suction flange) and the vapor pressure of that liquid is called the *net positive suction head available* (NPSH$_a$).

When the available NPSH is equal to or greater than the required NPSH, the pump will not cavitate. However, when the NPSH$_a$ is less than that required by the pump, cavitation occurs (Fig. 3.2). When cavitation occurs, the resulting vapor bubbles will tend to form at the area of lowest pressure, the impeller inlet. The bubbles are then caught up and swept outward along the leading face of the following impeller vane. Near the impeller tip, the pressure, due to losses, may once again be higher than the vapor pressure of the liquid, thus causing the bubbles to collapse. When these bubbles collapse, the energy released, known as the *latent heat of vaporization,* generates noise and may severely damage the impeller. Depending on the severity of the cavitation, a falling off of pump performance may also be noted.

3.1.2   NPSH testing—Too little, too late

The NPSH$_r$ by a pump at various flows and speeds is normally determined by testing in accordance with procedures found in *Hydraulic Institute Standards.*[1] Testing may be accomplished by either valve suppression or a vacuum method. In the valve suppression method of testing, a valve, located in the suction line leading to the pump, is used to reduce the suction pressure by throttling, thus creating varying NPSH$_a$ conditions. An alternate to valve suppression testing is to create a vacuum on the suction side of the pump (Fig. 3.3). This is done by using a tank or reservoir in which the pressure above the liquid is

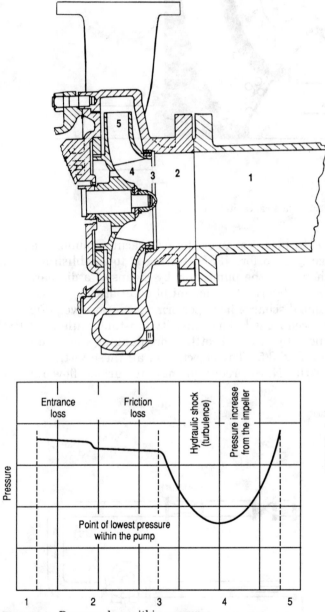

Figure 3.1   Pressure drop within a pump.

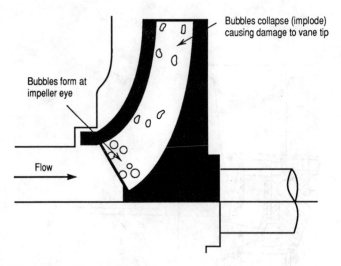

Figure 3.2   Classical cavitation.

reduced by means of an ejector or a vacuum pump. The procedure most commonly used for NPSH$_r$ testing is to establish a constant NPSH$_a$ and then vary the pump flow by means of a discharge control valve until a predetermined amount of deterioration (usually 3 percent) in the pump discharge head performance is observed (Fig. 3.4). An alternate procedure is to hold capacity constant while the NPSH available is reduced by either throttling or changing the vacuum on the pump suction (Fig. 3.5). The objective of all these testing methods is to establish the NPSH requirements, at various flow rates, for a given

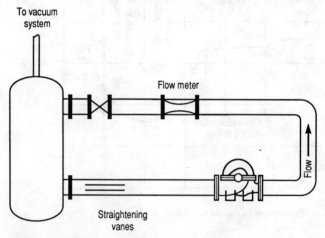

Figure 3.3   NPSH$_r$ testing—vacuum method.

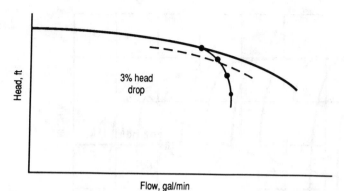

Figure 3.4 NPSH$_r$ testing—variable flow with constant NPSH$_a$.

impeller-casing combination and to construct an NPSH$_r$ versus flow capacity curve (Fig. 3.6). However, regardless of the method used, each testing procedure described determines the NPSH required by first detecting a deterioration in discharge performance. This means that cavitation had to have already been occurring, at some higher-flow condition, *before* the performance loss was noticed.

3.1.3 NPSH required versus available— A need for a margin

The pump manufacturer usually determines the NPSH required by testing, normally using water. These tests frequently are made on only the first casting(s) for a specific case or impeller pattern; subsequent

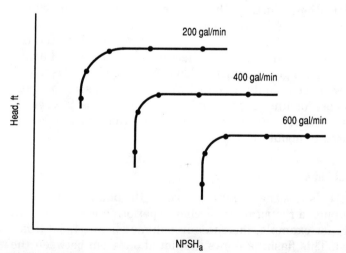

Figure 3.5 NPSH$_r$ testing—constant capacity with variable NPSH$_a$.

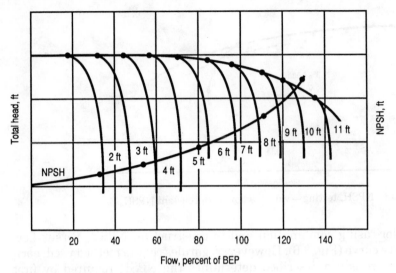

Figure 3.6 Constructing an NPSH$_r$ curve.

verification of the original test results often is obtained only when a purchaser specifies that an NPSH$_r$ test be made on a pump prior to shipment. However, experience has shown that variations in the NP-SH$_r$ curves may occur between pumps of the same size and model equipped with the same impeller and case patterns. This variation can be in excess of 5 ft, particularly on a small-capacity pump with a narrow impeller. These variations are a result of impeller eye area differences including vane angles, surface finish, and suction nozzle passage area variations. Further, the required NPSH will change for a pump in service as internal surfaces and clearances deteriorate. Thus, it is recommended that a reasonable margin be provided between the NPSH required, as quoted by the manufacturer, and the NPSH available over the entire range of predicted flows. A reasonable margin of 5 ft, at rated flow, is commonly accepted by many end users for most services. However, for known problem applications, such as vacuum tower bottoms, solvents, and boiler water feed, this margin is frequently increased to about 10 ft.

3.1.4 NPSH problems

When the NPSH$_a$ is less than that required by the pump, many problems other than just a reduction in discharge performance may result. Water, at 70°F, for example, increases in volume about 54,000 times when vaporized. This flashing blows the liquid out from between the impeller vanes. Excessive vibration can occur when some of these passages are filled with vapor while others contain liquid. Cavitation-induced pitting and erosion of internal parts may significantly reduce the life expectancy of the pump. As explained above, it takes energy to form vapor bubbles within a liquid. As these bubbles pass through the impeller to regions of higher pressure, the bubbles collapse. Adjacent walls and impeller vanes are subjected to tremendous shocks as the vapor gives up this energy and liquid rushes in to fill the voids left by the collapsing bubbles. This energy impacting on the surface of the pump effectively "blasts" off metal and severely erodes the surfaces.

Cavitation damage to a centrifugal pump may range from minor pitting to catastrophic failure depending on the pumped-fluid characteristics, energy level, and duration. However, most of the damage usually occurs within the impeller, specifically, the leading face of the vane toward the outer tip. This is the area where the bubbles will normally begin to collapse, releasing their energy onto the vane surface. The result is often a rough surface that feels like a cat's tongue. However, if this surface texture is found elsewhere on the impeller, i.e., on the back of the vane, on the shrouds, or in the inlet area, then the most likely cause is recirculation or some other mechanism other than classical cavitation.

A pump operating with insufficient $NPSH_a$ will often pump in spurts. When the pump is started, the liquid in the pump suction nozzle accelerates until the flow rate reaches the capacity at which it is to operate. As the fluid accelerates, the friction losses and pressure drop increase, thereby lowering the pressure in the suction nozzle until the liquid vaporizes. As soon as this occurs, the pumping action is reduced and flow decreases. With the decreased flow, the friction losses and pressure drop are less. The liquid no longer flashes, and the pump begins to regain performance. This increases the flow and restarts the entire cycle. Continued operation in this cyclic mode may eventually result in the failure of the mechanical seal and/or the bearings.

3.1.5 Short of $NPSH_a$, what next?

A centrifugal pump in the field that is cavitating often will sound as if it is pumping rocks, and frequently the discharge pressure will pulsate. A simple way to determine if the problem is a flow-induced NPSH problem is to slowly shut down on the discharge block valve. If the problem is flow-induced, the noise and the pulsations should go away as the flow is reduced through the pump. But if it is flow-induced, what can be done about it? Assuming that the higher flow through the pump is required for the process, there are several methods that can be employed to help solve the problem:

1. Check the level in the suction vessel or drum. If possible, raise the level, thus increasing the $NPSH_a$ to the pump.

2. Check the suction pressure. Ensure that all suction valves are open and that any suction screens or strainers are clean. If access is possible, ensure that the suction nozzle or vortex breaker from the suction vessel is free of debris.
3. Check the suction temperature. In many process applications, the liquid may be subcooled before it enters either the suction vessel or the pump. Try cooling the liquid some.
4. Put the spare pump on-line in parallel with the main pump. This will reduce the flow to each pump, thus reducing the $NPSH_r$ to each. If operation in this mode is required on a continuous basis, consider one of the following:

   a. Add a third identical pump, normally running two.
   b. Install new, higher-capacity pumps that require less NPSH, e.g., using larger eye areas, using double suction impellers, or operating at a slower speed.
   c. Install a pump equipped with a flow inducer. *Caution:* While inducers may reduce the $NPSH_r$ for a specific flow, some designs can significantly reduce the range of stable operation.
   d. Effectively lower the pump, not by putting it in a pit, but by installing vertical pumps which have the impeller inlet located several feet below grade.

3.2 Vortexing

Another hydraulic phenomenon that produces problems similar to those encountered when the $NPSH_a$ is less than required is called *vortexing*. Vortexing is a swirling, tornadolike funneling action in a liquid that occurs between the surface of the liquid and the draw-off nozzle in a bathtub, vessel, or vertical pump (Fig. 3.7). When vortexing occurs, air or vapor can be drawn into a pump, thereby effectively reducing the NPSH available. One volume percent of gas in the liquid will produce a noticeable drop in discharge performance, and 2 volume percent can result in as much as a 10 percent loss in pump capacity.

3.2.1 Submergence is not NPSH

Submergence and NPSH are not the same. It is possible to have adequate submergence and insufficient NPSH or vice versa. *Submergence* may be simply defined as the height of liquid above a vertical pump suction bell or submerged draw-off nozzle necessary to prevent the formation of a surface vortex at a given flow rate. A surface vortex, once completely formed, will allow air or gas to be pulled directly into the suction pipe or end bell. This can cause extensive damage to the pump and/or piping system as a result of imbalance-induced vibration,

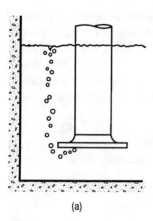

(a)

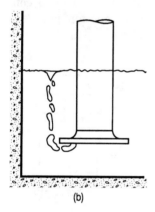

(b)

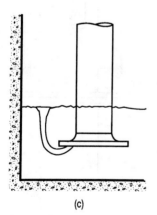

(c)

Figure 3.7   Vortexing:
(*a*) Type I—inception;
(*b*) Type II—performance
may start to deteriorate;
(*c*) Type III—breaks suction.

rubs due to running dry, and shock loading from potential hydraulic pulsations. The *Hydraulic Institute Standards* provide recommended practices and dimensions for sump designs for both vertical wet-pit pumps and side-intake applications. All proposed installations should be reviewed to ensure that there is adequate submergence *and* available NPSH over the entire range of expected pump operation.

Vertical wet-pit pumps   A vertical centrifugal pump (Fig. 3.8), has no suction piping and requires a minimum submergence level to prevent vortex formation. The pumps should have a properly designed sump with adequate distances between the end bell and the sidewalls and sump bottom. Frequently, to prevent surface vortexing, baffles and submerged or floating "rafts" are used. Vortex formation is unlikely when a vertical "canned" pump (Fig. 3.9) is used since, with this configuration, the barrel or can serves as an annular suction pipe (see Chap. 9 for further discussion of vertical pumps).

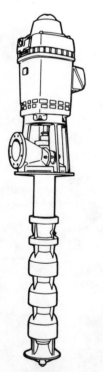

Figure 3.8   Vertical wet pit type pump.

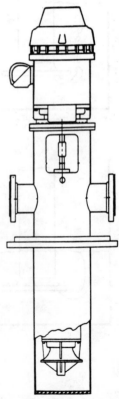

Figure 3.9   Vertical canned type pump.

Side-intake designs   The external, horizontal cooling tower circulating pump (Fig. 3.10) is a typical example of an installation utilizing a side-intake arrangement. The water level in the cooling tower basin determines the basin's height and therefore the cost. Consequently, a minimum water level is preferred. However, the level must be sufficient to meet the pump's $NSPH_r$ as well as to prevent vortexing. To accomplish this, a baffle or spoonbill is frequently employed.

Design of the suction piping is also critical. The eccentric reducers in the suction line must be installed with the flat side up in order to avoid an air pocket. Butterfly valves that offer little flow resistance and no air pockets in the housing are used in lieu of standard gate-type valves.

Typical process pumps   Vortexing may be a problem with any typical centrifugal pump application. It may develop as a result of a very low liquid level in a suction vessel or when the fluid velocity in the suction piping becomes too high. To help counter this phenomenon, mechanical vortex breakers are sometimes used where the suction piping exits the vessel. However, these devices are frequently overlooked during maintenance and may be inadvertently damaged, plugged, or left out.

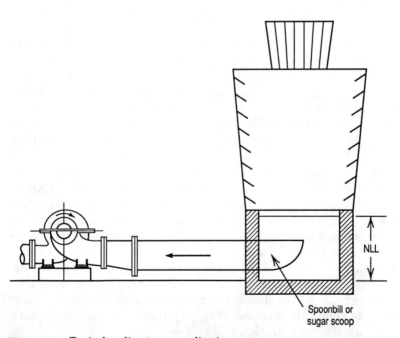

Figure 3.10 Typical cooling tower application.

3.3   Recirculation, a Neglected Factor

While published performance curves often show smooth, stable operating characteristics from zero to well past the BEP flow capacity (runout flow), many pumps cannot function over these wide ranges without incurring severe damage. In fact, many pumps cannot be operated below 50 to 60 percent of BEP flow without incurring severe hydraulic instabilities. The cause of these problems is internal recirculation, a hydraulic phenomenon related to the suction specific speed, $NPSH_r$, efficiency, and range of the pump. Fluid recirculation can occur at the impeller inlet and outlet. It generally occurs during operation below the BEP flow rate but frequently occurs when the flow rate exceeds BEP.

A centrifugal pump is designed to operate at its best efficiency at only one flow rate at a fixed rotational speed. Liquid flow patterns in the case and within the impeller internal channels are very complex, especially at off-design conditions. Most pump handbooks assume uniform flow, although the actual flow seldom approximates these conditions. If the system requirements call for capacities that differ significantly from the design point, the pump will have to operate at some lesser efficiency. The drop in efficiency is due to increased internal losses resulting from recirculation.

The pump driver has to provide more energy to overcome these losses, energy that goes into producing heat within the pumped fluid. Specifically, the heat generated is a result of friction and viscous shearing within the fluid. Changes in friction losses are relatively constant with changes in capacity. However, shear losses, because of the many different directions and velocities that the fluid assumes within the pump, increase significantly as capacity is reduced from the BEP. Effectively, the fluid enters the eye of the impeller but has insufficient energy to pass through it. Therefore, the fluid begins to fold back on itself, forming eddy currents and recirculation flow patterns (Fig. 3.11).

During recirculation, the internal flow is unstable, with violent changes sometimes occurring from channel to channel. Stall, backflow, eddy-type circulation, turbulence, and cavitation are all mechanisms that can and do take place. Frequently, when severe recirculation occurs, loud popping or banging noises are heard at random frequencies coming from the pump. In addition, random hydraulic pulsations may be observed in the suction piping up to several pipe diameters upstream of the pump suction. The intensity of these noises and pulsations will increase as the flow is further reduced. Sustained operation in this mode should be avoided as it may result in significant pump and/or piping system vibration and failure.

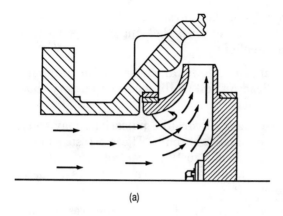

(a)

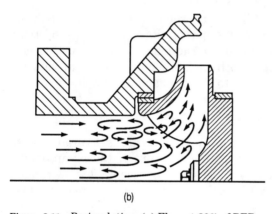

(b)

Figure 3.11   Recirculation: (*a*) Flow at 80% of BEP—
little if any recirculation or blockage;
(*b*) flow at 30% of BEP—heavy recirculation and almost complete blockage.

3.3.1 What are the effects of recirculation?

There are several destructive effects of recirculation:

1. *Flow recirculation at the impeller eye:*   This is generally found during off-design flows. Damage caused by this flow instability is found
   at the impeller eye or inlet areas of the casing.

2. *Flow recirculation at the impeller vane tips:*   Damage occurs at the
   impeller outside diameter. This eddy flow is caused by reduced capacity flows and improper impeller tip clearance or alignment (gap
   A and gap *B*).

3. *Flow recirculation around impeller shrouds:* Damage is seldom seen on the impeller; it is seen as thrust bearing damage, particularly on double suction impellers.

4. *Impeller erosion:* The turbulent flows can cause erosion of the impeller vanes on both the leading and trailing edges in a very short time.

5. *Impeller failure:* Low-frequency hydraulic pulsations may cause fracture-type failure of the impeller shrouds. These pressure pulsations may be in the magnitude of 5 to 10 percent of the total head and may cause an impeller failure after only a few hours of operation at extremely low flows.

6. *High failure rate of mechanical seals:* The hydraulic pressure pulsations are very destructive to mechanical seals and may cause opening of the primary sealing faces.

7. *High bearing failure rate:* The pump rotor is moved by the hydraulic pulsations and can cause an impact failure of bearings, particularly ball thrust bearings. In double suction pumps, the suction side pulsations are out of phase with each other and can be at varying frequencies, causing the rotor to shuttle back and forth.

3.3.2 Hydraulic stability guidelines

Because recirculation, with the associated unstable hydraulic flow characteristics, can result in mechanical failure of the pump, traditional pump curves—especially those showing acceptable flow ranges from shutoff to greater than 120 percent of BEP flow—are often misleading or even invalid.

Similar to the discharge specific-speed term, which is used to describe geometrically similar impeller discharge designs, the *suction specific speed* is used to describe geometrically similar impeller suction designs. However, it also can be used to describe the performance of a pump with regard to its range of stable hydraulic operation and susceptibility to recirculation. Suction specific speed N_{ss} is defined by

$$N_{ss} = \frac{(r/min)[(gal/min)/eye]^{1/2}}{(NPSH_r)^{3/4}} \tag{3.1}$$

Values for capacity and NPSH required are taken at the pump's best efficiency point for the design impeller (usually the maximum diameter available). In case of double suction impellers, which actually have two inlets, or two eyes, the suction specific speed is based on the performance *per inlet,* or *per eye,* so that the total capacity should be divided by 2.

Failure to understand the problems of recirculation can lead to severe maintenance and operating troubles that can be directly traced to high suction specific-speed problems. J. L. Hallam conducted a study of several hundred pump failures (480 pumps and 1881 failures) that had occurred in the refining industry over 5 years. He concluded that pumps with high suction specific speeds (above 11,000) had a failure rate approximately double that of the ones with lower ratings (Fig. 3.12).[2]

There is no industrywide consensus on the N_{ss} value for a conservative versus a marginal centrifugal pump design. However, many pump maintenance and application engineers have found the following guidelines to be reasonable:

1. For cold-water and general service applications, suction specific speeds of 8500 and lower are selected.

2. For boiler feed and condensate applications and for general hydrocarbon service, values of suction specific speeds typically range between 8500 and 11,000.

3. Pumps designed for suction specific speeds in excess of 12,000 are generally for special applications only.

The severity of the problems associated with recirculation is partially dependent on the available energy. Small-sized, single-stage overhung pumps, requiring relatively low horsepower [<250 horsepower (hp)], generally are not adversely affected by recirculation; therefore, higher values of N_{ss} can be used. In addition, some pump

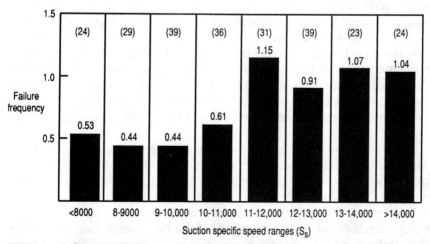

Figure 3.12 Suction specific speed versus failure rate. Note: Number of pumps in each range shown in parentheses.

designs significantly reduce the effects of recirculation by utilizing internal channels to divert the recirculating flow away from the impeller inlet. However, before any pump with a suction specific speed in excess of 12,000 is used, an acceptance test that covers the whole operating range of the pump should be carried out.

3.3.3 Remedies for existing recirculation

The following are a few practical considerations that should be understood in order to head off potential recirculation troubles:

1. A high suction specific-speed impeller, one that has a low $NPSH_r$, is very susceptible to recirculation. Calculation of suction specific speed is about the only predictive measure available to avoid potential trouble from undesirable, low-flow recirculation.

2. Flow disturbances are more common with double suction pumps, thus making them more vulnerable to recirculation than single suction pumps. This characteristic is due, in part, to the potential for uneven suction flows to the impeller eyes. This uneven flow may be the result of the orientation of external suction valves and strainers and/or differences in the suction flow channels resulting from core shifts and other casting problems.

3. Really serious recirculation problems, because of the energy levels involved, are usually limited to high-energy pumps (greater than 650 ft of head and 250 hp per stage), However, even small pumps can be damaged or suffer performance losses as a result of recirculation.

4. To suppress the destructive characteristics of recirculation, a very high $NPSH_a$ may be necessary. As a pump is operated back on the curve, eddy currents begin to form in the eye of the impeller and in the suction casing, partially reducing the cross-sectional area of the suction nozzle. However, since at a given flow there is a constant mass flow through the pump, the fluid velocity through the inlet channels must increase and as the velocity increases, the pressure drop due to friction and shear also increases. If the drop is large enough to go below the vapor pressure of the liquid, the pump will begin to cavitate. This tendency may be suppressed by raising the available suction head.

5. Recirculation is much more damaging with some liquids than with others. Pure liquids, such as water, are homogeneous with vaporization occurring instantaneously at a single temperature for a given pressure. In addition, liquids like water have a high vapor-to-liquid volume ratio. When vaporization occurs within a pump, this characteristic tends to "blow" the liquid out of the impeller. A fluid consisting of a mixture or fractions of different liquids will vaporize over a range

of temperatures at a given pressure. The wider the "boiling range" and the smaller the vapor-to-liquid volume ratio, the less violent the cavitation reaction tends to be.

6. In multistage pumps, suction recirculation normally occurs only in the suction or first stage. However, exit tip recirculation may occur at any stage.

7. If the pump is already installed, most of the corrective items discussed above are not available; other remedies are needed. The most effective remedy is to increase the fluid flow through the pump such that it always operates at or near BEP flow. For some systems, this may require the installation of a bypass line around the pump and a control valve. The first costs of installing the bypass line and the cost of pumping the excess flow may seem high. However, for very high suction specific-speed pumps, the potential savings in maintenance costs and improved reliability are generally sufficient to justify the capital outlay.

3.4   Minimum Flow Requirements

Operating a centrifugal pump at less than the design or BEP flow rate increases the potential for problems. But what should the minimum flow limits be?

Traditional pump curves usually indicate stable flow from zero (commonly called shutoff) to well past the best efficiency point (runout flow) with no hint of any potential problems. Normally provided with these curves is an NPSH$_r$ versus capacity curve that indicates a steady increase in the required suction head over the same range. Figure 3.13 is a typical head and NPSH$_r$ versus flow performance curve for a centrifugal pump. It is similar to thousands of other curves found in most any pump vendor's catalog.

In the past, when asked if a pump had a minimum flow restriction, most vendors responded no or at best maybe some minimal flow (usually based on potential heat rise), such as 5 to 20 percent of BEP. In reality, the lower 50 to 60 percent of these operating curves for many pumps is invalid and cannot be relied upon to predict future pump operations under actual field conditions. Today, most pump vendors and users generally agree that establishing the correct minimum flow requirements for a centrifugal pump is critical for its reliable operation. They also now recognize that the magnitude of the minimum required flow rate must take into consideration such things as normal wear, NPSH$_r$, vortexing, suction and impeller tip recirculation (gaps A and B and N_{ss}), characteristics of the fluid being pumped, available energy levels, and the potential for excessive heat rise.

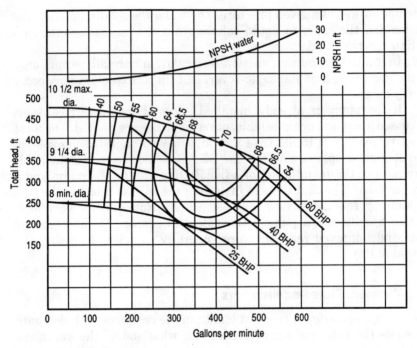

Figure 3.13 Typical pump vendor performance curve.

3.4.1 Temperature rise

Limitation of the heat generated at low-flow conditions is probably the most widely used criterion for determining the minimum flow for a centrifugal pump. The temperature rise of the liquid within a pump is basically a function of the pump inefficiency. The maximum temperature rise allowed must be limited to that which will not excessively distort the case, causing seizure, or to that which will cause flashing of the pumped liquid.

For single-stage pumps this normally occurs at the inlet to the impeller. However, for multistage pumps, the design engineers must also look at the difference between the discharge temperature and the flash point of the liquid (at suction pressure) seen within the balancing cavity. If, however, when taking this into consideration, the engineer looks at only rated flow, then pump failure may still occur. During the start-up of a multistage standby pump, before the pump comes up to speed and begins to develop sufficient pressure to open the discharge check valve, there is no flow through the pump. There is a rapid rise in the pumped fluid temperature that causes flashing across the axial balancing drum. As a minimum, this may result in the failure of the thrust bearing or the catastrophic failure of the pump due to seizure.

To prevent this, a minimum flow recirculation line extending from the discharge piping (upstream of the discharge check valve) back to the suction vessel must be provided. The line should be sized to allow sufficient flow through the pump to prevent flashing during start-up.

For single-stage pumps, the maximum temperature rise to prevent flashing can be calculated, at various flow points, by the following relationships:

$$\text{Temperature rise} = \frac{H}{778E} \qquad (3.2)$$

where temperature rise = allowable rise, °F
H = total head, ft
E = pump efficiency

$$\text{Maximum temperature rise} = T_{\text{sat}} - T_{\text{inlet}} \qquad (3.3)$$

where T_{sat} = flash temperature at-suction pressure.

Note: For multistage pumps, the temperature rise can be estimated by using the overall head rise and pump efficiency.

For a more accurate calculation, individual stage heads and efficiencies are required. For some special pumps, operating in extremely high or low (cryogenic) temperatures, the heat rise (or drop) may be limited to the thermal growth of the casing and/or impeller. If the thermal growth rate of the casing material is significantly different from that of the impeller, there is the potential that one component may grow into the other during start-up or shutdown, resulting in a minor rub or, at worst, a catastrophic shaft seizure. For some chemical process pumps the maximum temperature rise is limited not to vaporization or thermal growth, but by that temperature necessary for some chemical reaction to take place, e.g., polymerization or solidification.

3.4.2 Hydraulic design

The proper minimum flow requirement for most centrifugal pumps should be based on the hydraulic design. Commonly referred to as the *minimum continuous stable flow,* this flow rate is normally higher than that based on only the temperature rise. Minimum flow requirements, based on hydraulic design, are given in Sec. 3.3.2.

The minimum continuous stable flow required tends to decrease as the pump energy level decreases. For low-flow, low-head (low-horse-power) pumps, even with relatively high suction specific-speed impellers, recirculation is seldom a significant problem. Quite frequently, systems requiring low-flow applications have an abundant NPSH_a, sufficient to suppress any recirculation. However, even in systems having less available suction head, while inlet eddy currents may still form, the energy available is usually too little to be destructive or to significantly reduce performance during low-flow operations.

For high-energy pumps, the opposite is normally true. The large pressure differentials and the energy concentration inherent to most high-energy pumps tend to make them more susceptible to damage and performance impairment, as a result of even minor recirculation. Therefore, the minimum stable flow for these pumps is usually closer to the BEP flow.

3.4.3 Characteristics of the fluid being pumped

The minimum flow may also be theoretically reduced when mixtures of hydrocarbons or other liquids are pumped. This is because at low flows, where inlet vane recirculation can cause local vaporization, the energy released is usually much lower than that with a fluid such as cold water. Therefore, minimum continuous stable flow usually is lower based on the vibration and noise criteria.

3.4.4 Other factors

The designing and casting of an impeller and casing are much of the "black art" of pump design. The form of the single and double volute case may cause an uneven distribution of pressure around the periphery of the impeller, producing a radial load perpendicular to the shaft axis (Fig. 3.14). The magnitude of the load varies directly with the impeller width, diameter, and head developed. The greatest radial load occurs at low flows and is minimal at the BEP flow rate (Fig. 3.15). This radial load imbalance increases bearing loads and shaft deflections that could lead to premature seal, shaft, or bearing failure. Shaft

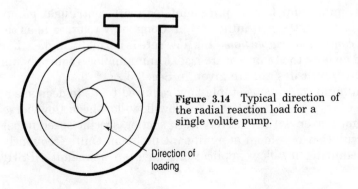

Figure 3.14 Typical direction of the radial reaction load for a single volute pump.

Direction of loading

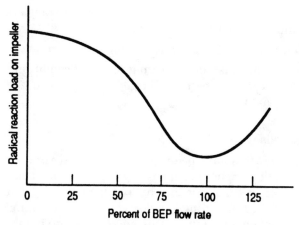

Figure 3.15   Radial reaction load versus flow for a single volute pump.

deflection due to radial reaction can be predicted quite accurately and therefore should be taken into consideration when the minimum flow requirements are specified. In a well-designed pump, radial load sets the minimum flow in the same range as the thermal rise, or about 10 to 20 percent of BEP.

3.4.5 Setting minimum flow

In addition to the many items discussed above, there are a number of practical considerations that must be understood when minimum flows are set.

1. The selection of a suitable minimum flow level for a pump depends heavily on the time factor. Is the low-flow condition almost continuous, a momentary emergency situation, or something in between? Such time requirements should be incorporated into the specified operational performance.

2. Recirculation cavitation in the impeller inlet is most likely to occur in pumps designed for lowest $NPSH_r$ (high suction speed); therefore these impeller designs require higher minimum flows.

3. Recirculation is much more damaging with some liquids than with others. In water, vaporization can occur instantaneously, so the minimum flow must be higher.

4. Double suction pumps are more vulnerable to recirculation than single suction ones and may require minimum flows in the BEP range of 60 to 70 percent.

5. Energy levels of 600 to 650 ft of head and 250 to 300 hp per stage are usually the lower limits of really serious instability.

6. NPSH available is a major factor in determining minimum flows.

7. To protect the pumps from instability during low-flow conditions, bypass systems are necessary in many larger pumps, assuming speed reduction is not available.

3.4.6 Minimum flow guidelines

For many years pump operators have been cautioned against running pumps below a minimum flow capacity, supposedly because of heat rise within the pump. However, it turns out that hydraulic instability, not heat rise, is the limiting factor. Avoiding low-flow conditions goes hand in hand with the proper sizing of a pump. During application, pumps should not be oversized, in either capacity or head. The recommendation is to select the pump as close to the best efficiency point as possible while keeping reserve head and capacity requirements to a minimum.

Experience is the major factor in establishing minimum flows. Here are some simple guidelines for estimating minimum flows:

1. Assume that no pump is designed to operate for longer than about 15 min below 50 percent of BEP flow, and make this your baseline.

2. Provide minimum flow bypass piping for high-energy pumps. The bypass line should be piped to permit heat dissipation, i.e., routed through a heat exchanger or back to the suction vessel if it is of sufficient volume for thermal dilution.

3. Impeller patterns with suction specific speeds greater than about 11,000 require a minimum flow of 60 to 70 percent of BEP.

4. A double suction impeller will also require minimum flows of 60 to 70 percent of BEP.

5. A pump with a suction specific speed of 20,000 may require a minimum flow of 100 percent since the pump is stable only near BEP.

6. Correction factors that reduce the minimum flows in the handling of hydrocarbons should be used carefully. It is true that a mixed chemical or petroleum liquid is composed of fractions that vaporize at different temperatures. The cavitation that occurs during recirculation is of a mechanical origin and is not totally temperature-related. Also, some narrow-boiling-range hydrocarbons tend to react as water does.

Any single percentage number for minimum flow can only be arbitrary. If the pump is expected to run at other than design flows, each situation should be spelled out as accurately as possible by the individual specifying the pump. By providing the vendor with data that are as complete as possible and applying some of the guidelines discussed above, the pump operator can achieve a pumping system that can be relied upon for continuous stable operation in low-flow applications.

References

1. *Hydraulic Institute Standards,* 14th ed., Hydraulic Institute, Cleveland, Ohio, 1983.
2. Jerry L. Hallam, "Centrifugal Pumps: Which Suction Specific Speeds Are Acceptable," *Hydrocarbon Processing,* April 1982.

4

Hydraulic Loads

4.1 Introduction

When centrifugal pumps are operating, hydraulic loads or forces are developed on the impeller and shaft. These hydraulic loads are a combination of both radial and axial forces that act on the impeller, shaft, and the bearing system. Providing the necessary mechanical resistance to shaft deflection and displacement from these hydraulic forces encompasses the mechanical design of the pump.

4.2 Radial Hydraulic Loads—Volute Design

The design of the casing that surrounds the impeller determines the unbalanced radial load which is exerted on the impeller and, in turn, on the rotor and bearing assemblies. The pump designer tries to size the volute cross-sectional area such that, at design flow, the fluid velocity is nearly constant at all points across the volute area. However, when the pump is operated at capacities lower or higher than the design, this velocity is no longer constant, and thus an unbalanced load or radial reaction force is created on the impeller. This radial force is perpendicular to the shaft axis and will increase from the tongue, or cutwater, around to the throat of the casing, as shown in Figs. 3.14 and 3.15. At about 50 and 120 percent of BEP flow, the direction of the deflection is 90° to the cutwater or volute tongue axis.

In double volute casings (Fig. 4.1), a "false wall" or second volute is cast into the casing to divert a portion of the fluid directly to the volute area of the pump such that it cannot react on the impeller. Hence, the unbalanced forces around the impeller tend to be more equal and counteracting in their effects on the impeller. The net unbalanced load is much lower than that found in a single volute casing (Fig. 3.14). Most

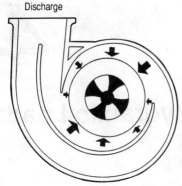

Discharge

Figure 4.1   Radial hydraulic loads—double volute, single-stage pump.

handbooks leave you with the impression that since double volute pumps have cutwaters diametrically opposed, the radial imbalance is in directly opposite directions at any given time and therefore balances. Because the volutes are cast into the case and little, if any, machine work is done on them, they tend to have uneven surfaces and are irregular in shape. Complete hydraulic balancing does not occur; however, the radial reaction forces are significantly reduced by the double volute.

4.3   Radial Hydraulic Loads— Diffuser Designs

Diffuser-type cases have evenly spaced, symmetric minivolutes, surrounding the entire impeller, and result in a more complete balance of the radial reaction forces (Fig. 4.2). In order to maintain this balance, it is important that the fluid exiting from the diffuser passages be rapidly and uniformly collected and removed. Otherwise, hydraulic

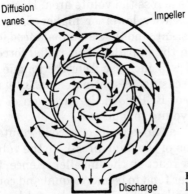

Diffusion vanes

Impeller

Discharge

Figure 4.2   Radial hydraulic loads—diffuser designs.

shock and recirculation losses can cause a radial imbalance to be set up external to the diffuser, which will react on the impeller. It is for this reason that symmetric return channel vanes are located adjacent to diffuser rings in multistage pumps to lead the flow from the diffusers uniformly into the inlet of the next stage impeller. In the last stage of a diffuser pump, it is common practice to collect the flow in a chamber axially away from the diffuser ring itself, so that imbalance forces will not be distributed to the impeller.

4.4 Radial Clearances—Impeller to Casing

To prevent excessive deflection at critical running clearances and critical components, such as mechanical seals and bearings, the radial hydraulic load (P) and shaft deflection should be taken into consideration when sizing the shaft (Fig. 4.3). The resulting reaction forces (R_1 and R_2) can cause shaft deflections of a magnitude greater than the wearing ring clearances. This results in the potentially rapid deterioration of the wearing rings, bearing failures, and shaft breakage due to fatigue failure of the shaft material. Reduced impeller radial gap, combined with poor casting quality and recirculation at low flows, is the most frequent cause of impeller and diffuser volute tip breakage. Many engineers believe that reduction of the normal radial gap (gap "B," shown in Fig. 4.4) between impeller and diffuser/volute improves efficiency. Not only is this incorrect, but also the reduced radial gap can cause impeller structural failures.

As an impeller vane passes a stationary vane (diffuser, volute

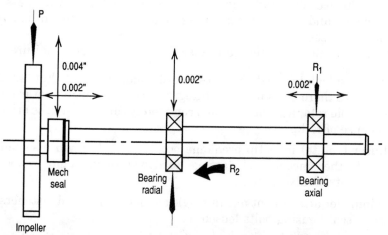

Figure 4.3 Shaft bending loads—hydraulic and mechanical factors, cantilever pump design.

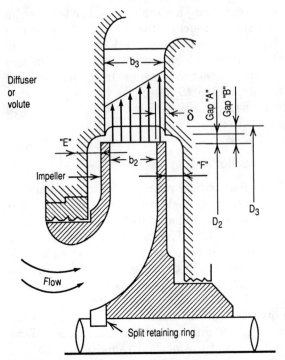

Figure 4.4 Nomenclature of impeller and diffuser geometry radial gaps.

tongue, or cutwater), a hydraulic shock occurs. This can be observed in the liquid pressure fluctuations or noticed on the rotor vibration. The distinct influence of the radial gap on pressure pulsation at vane passing frequency and on the radial forces generated is shown in many vibration analyses.

This vane passing vibration can cause the following types of failures:

1. Impeller sideplate breakage is the most common result (Fig. 4.5). Sometimes there is complete disintegration of the impeller, or even of the whole rotor assembly, which completely camouflages the original cause of failure.

2. Diffuser (volute) inlet-tip breakage occurs (Fig. 4.6) and also can cause further mechanical damage to downstream pump components, control valves, etc.

3. The impeller attachment nut in single suction pumps and fasteners such as seal housing bolts loosen.

4. Secondary piping components break, such as a seal injection line, cooling water, etc., due to pressure pulsations.

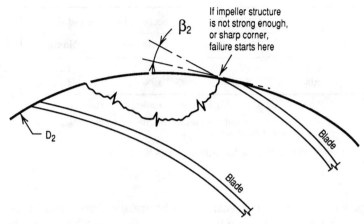

Figure 4.5 Fatigue failures of impeller due to pressure pulsations from improper radial clearances.

5. The shaft breaks, generally outboard of the impeller in double suction designs.

6. The thrust bearing fails, especially in double suction types which are designed for very low-thrust loads.

7. High-pressure pulsations possibly can cause malfunctioning of the control system.

8. Mechanical seal failures occur.

4.5 Corrective Actions—Vane Passing Frequency

Dr. Elemer Makay, working with the Electric Power Research Institute (EPRI), has determined that careful machining of the volute or diffuser tips to increase gap *"B"* while maintaining gap *"A"* can reduce

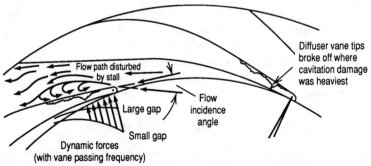

Figure 4.6 Diffuser (volute) failures due to pressure pulsations from improper radial clearances.

Type pump design	Gap "A"	Gap "B"*—Percentage of impeller radius		
		Minimum	Preferred	Maximum
Diffuser	50 mils	4%	6%	12%
Volute	50 mils	6%	10%	12%

*$B = 100(R^3 - R^2)/R^2$; R^3 = radius of diffuser or volute inlet; R^2 = radius of impeller.

NOTE: If the number of impeller vanes and the number of diffuser/volute vanes are both even, the radial gap must be larger by about 4.

Figure 4.7 Recommended radial gaps for pumps—impeller to casing.[1]

the vane passing frequency greatly. The hydraulic forces can be reduced 80 to 85 percent in a pump by increasing the gap. The larger the gap, the more the flow can smooth out before it encounters a diffuser vane or volute tongue. A diametral gap of about 6 percent of the impeller *radius* for a moderate head per stage impeller is widely accepted. For higher heads per stage (1600 ft or more), 8 percent or more is used. Values for volute pumps are usually in the 8 to 12 percent range, depending on the design, specific speed, and head per stage. If the minimum tip clearances are not present in the pump, it will be necessary to grind away part of the cutwater or casing volute to gain the clearance. The cutwater should not be too blunt but should be ground to about ³⁄₁₆-in maximum width. The cutwaters must be 180° apart on a double volute. There is no loss of overall pump efficiency when the diffuser or volute inlet tips are recessed. A small efficiency improvement results from the reduction of various energy-consuming phenomena: the high noise level, shock, and vibration generated with vane passing frequency as well as the stall generated at the diffuser inlet. Recommended dimensions for the radial gaps are given in Fig. 4.7.

The vane tip clearance, gap *"B,"* shown in Fig. 4.4 controls the strength and amplitude of hydraulic shock created at vane passing frequencies. The impeller shroud clearance, gap *A*, controls the severity of pressure pulsation behind the impeller hub and shroud which give rise to high axial direction dynamic forces with low frequencies. The phenomenon of recirculation at the discharge of an impeller can, under certain conditions, trigger recirculation at the suction.

In addition to increasing the radial gap, more effective and reliable impellers can be achieved by several design changes:

1. Use an odd number of impeller vanes with double volute pumps. Avoid an even number for double volute or even-numbered diffuser vanes.

2. Extend the impeller middle shroud plate of double suction impellers out to the outside diameter to reinforce the impeller structure.

3. Stagger the right and left side of the vanes of double suction impellers to reduce hydraulic shocks and alter the vane passing frequency.

4. Maintain proper *axial* clearances between the impeller exit and diffuser volute inlet.

5. Reduce *radial* gap A clearances to about 0.050 in between the shrouds and casing.

6. Increase the impeller sidewall thickness.

7. Impellers manufactured with blunt vane tips can cause trouble by generating hydraulic "hammer" even when the proper gap "B" distance from the cutwater. The blunt vane tips cause disturbances in the volute. The vane tips should be underfiled, as discussed under impeller trimming practices in Chap. 8.

4.6   Shaft Failures

Shaft failures may occur in double suction pumps having a large bearing span. One peculiar feature of these shaft failures is that, in a great majority of cases, the shafts fail immediately outboard the impeller where torque stresses do not exist. The weight of the coupling sets up a negative bending moment at the inboard bearing, which results in a greater bending moment and fatigue stress in the plane just outboard of the impeller than in the plane just inboard of the impeller. Frequently threads or other stress risers are located in this outboard plane, and failure occurs from the bending action. These bending stresses, coupled with excessive radial hydraulic forces, can cause frequent shaft breakage. An effort should be made to reduce both bending forces.

A readily available correction will often reduce the shaft bending stresses and breakage. If the overhung coupling weight is reduced, the bending problem is also greatly reduced. Many motor manufacturers oversize the coupling required, and by just reducing the coupling one size, 80 to 90 lb of weight can be eliminated.

4.7   Axial Thrust Problems in
Single-Stage Pumps

Axial loads occur in a pump because of axial pressure imbalance. In an overhung, single suction pump, suction pressure is exerted on the hub area of the impeller. This suction pressure force is resisted by atmospheric pressure acting over the same area. Thus, an imbalance will occur when suction pressure is either more or less than atmospheric pressure (Fig. 4.8). In a double suction pump with two bearings and the shaft extending through the eye of the impeller, this axial thrust due to suction pressure is about equal on each side unless there are flow disturbances.

A centrifugal pump impeller of conventional single suction design creates a thrust force on itself. This is due to the effect of the discharge pressure of the impeller acting behind the back shroud and behind the front shroud (Fig. 4.9). The area behind the front shroud is between the outer diameter and the diameter of its eye wearing ring. The same is true of the area behind the back shroud, but its wearing ring diameter is generally smaller. The area behind the back impeller shroud is then much larger. The net effect is a larger area on which the differential pressure developed by the impeller is acting.

In all pumps, residual unbalanced axial thrust must be resisted by the thrust bearings. It is important to be able to predict the magnitude of this unbalanced axial load in order to properly select and size the thrust bearings.

4.8 Balancing of Axial Thrust in Single-Stage Pumps

The amount of axial thrust developed in single-stage pumps is almost totally dependent on the impeller design. The amount of hydraulic balancing is also determined by the impeller design. The effects of

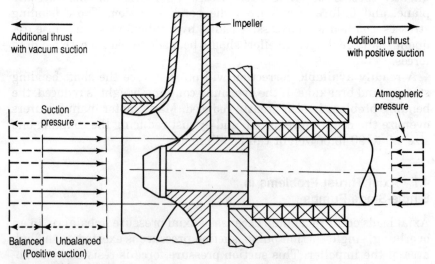

Figure 4.8 Effects of suction pressure on axial thrust of single-stage cantilevered pumps.

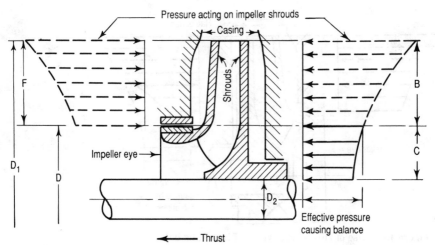

Figure 4.9 Effects of discharge pressure on axial thrust of single-stage cantilevered pumps.

suction pressure on axial thrust are shown in Fig. 4.8 and are common to all impeller designs. The effects of discharge pressure vary among the five basic impeller designs.

1. One design uses a closed-type impeller with a full back shroud and an impeller eye wear ring (Fig. 4.9). The area close to the shaft is open directly to discharge pressure, and the resultant pressure is quite high. However, the rotational effect of the liquid trapped between the impeller shrouds and the casing wall reduces the pressure somewhat, so that the pressure at the shaft and the wear ring is approximately 70 percent of the total head plus suction pressure.

2. Open impellers tend to generate more thrust action than an equal-diameter closed design. They also tend to be less efficient than closed-type designs because of greater leakage across the face of the vanes. The impeller position must be adjusted axially to maintain correct clearance.

3. On smaller pumps, complete balance at all capacities is not necessary because a large enough ball bearing can handle any residual axial force that exists. Radial ribs or vanes on the back shroud (Fig. 4.10) are sized to reduce the pressure on the back impeller shroud, as indicated by the shortened arrows. There is no change in the pressure on the front shroud on the impeller. This method gives axial balance at only one point on the head-capacity curve. This method is used with overhung impellers in ANSI pumps. API 610 specifications do not permit the use of ribs for thrust balancing. The design is generally not desirable because of high seal and bearing maintenance in the more severe service API pumps.

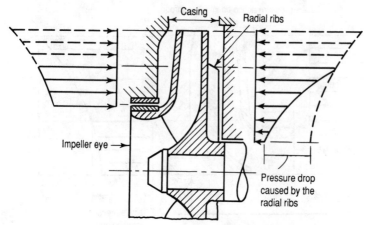

Figure 4.10 Use of ribs to achieve axial hydraulic balance.

Simplicity is supposed to be the chief advantage of the rib design. Actually the ribs complicate maintenance because the rib-to-casing clearance must be held under 0.025 to 0.030 in. The impeller must be repositioned axially as the ribs wear; the adjustment mechanism adds complexity to the pump stationary components. The stuffing box pressure is only slightly reduced from the 70 percent value of the impeller without the ribs.

4. Back wear rings and balance holes are a common method of achieving thrust balancing on an overhung pump (Fig. 4.11). Wear rings on the back shroud are connected to the enclosed chamber inside the

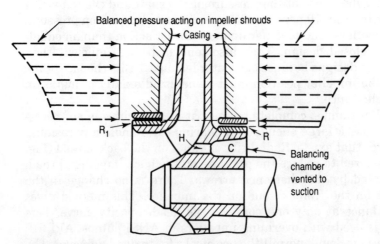

Figure 4.11 Use of back wear ring and balancing holes to achieve axial hydraulic balance.

ring by holes to the suction side of the impeller. Hydraulic balance is never attained for the pump's entire capacity range. The balance holes in each liquid channel allow suction pressure to exist on both sides of the impeller, up to the diameter of the back wear rings. Thus, the only imbalance which would occur is that which exists if the front and back wear rings are not exactly the same diameter. The back ring's diameter determines the amount and direction of the residual thrust.

If the shaft runs through the eye, then the back ring has about the same diameter as the front ring. If the impeller is overhung, high suction pressure may develop an appreciable thrust toward the back shroud. To balance the thrust under these conditions, the diameter of the back ring is reduced as necessary to compensate for the thrust due to suction pressure on the shaft's end.

As the wear ring clearance increases due to wear, the stuffing box pressure will go up. The balance hole areas should be at least twice the ring clearance area. If the wear is equal on both sets of rings, the originally balanced areas will still be balanced since the pressure drop will affect both alike. Since leakage through the back ring has increased, pressure may build up in the balancing chamber and create a thrust toward suction. As a result, the final effect of wear is dependent on the actual impeller design.

5. Another way to reduce axial thrust is to mount impellers back to back. In effect, that is what a double suction impeller is: two impellers cast as an integral unit. Liquid entering the pump divides and flows into the impeller from opposite ends. The most common example of this arrangement is the horizontally split double suction pump (Fig. 4.12). Under ideal conditions, the hydraulic axial forces acting on one side of the impeller balance those on the other, and there is no thrust. In actual operation, however, it develops because of a number of factors:

   a. When either side of the suction becomes partly obstructed, flow is no longer equally distributed between the sides of the impeller, and the hydraulic axial forces become unbalanced, as shown in Fig. 4.13.

   b. The casing casting is not symmetric on both sides of the impeller.

   c. A horizontal elbow at the suction will cause unsymmetric flow into the impeller eyes, as shown in Fig. 4.14.

   d. Recirculation caused by operating at reduced flows (below 60 percent of BEP) can cause axial hydraulic imbalance and, worse yet, an alternating thrust action which can destroy the ball bearings by impact loading.

      In large pumps, impeller eye areas are so great that 1 or 2 lb of difference in the suction pressures on the two sides can easily

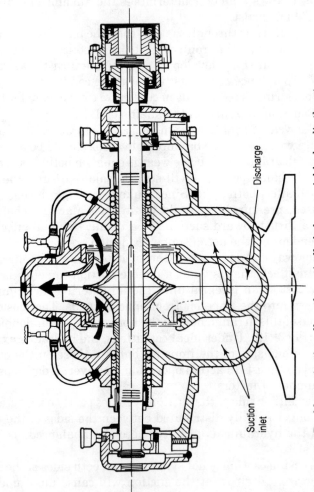

Figure 4.12 Double suction impellers theoretically balance axial hydraulic thrust.

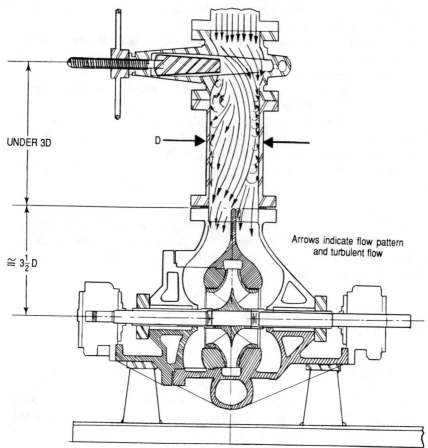

Figure 4.13 Effect of turbulent flow through a valve on double suction pump.

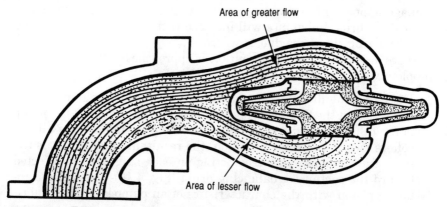

Figure 4.14 Effect of elbow in horizontal plane on suction flow to a double suction pump.

produce a thrust of 400 lb. Double suction pumps require thrust bearings of ample size.

If flow remains evenly divided between two sides of a double suction impeller, wear has little effect in throwing the hydraulic axial forces out of balance as long as both sets of wear ring clearances wear uniformly. If they do not wear similarly, more leakage flows through one wear ring than through the other, to cause an imbalance of the hydraulic axial forces. Pressure will be lowest on the shroud adjacent to the greatest leakage with a consequent thrust in that direction. Wear rings should have the same diameters and clearance to reduce the alternating thrust actions.

Complete axial hydraulic balance is never achieved in any pump design. Poor maintenance practices, such as unequal wear ring clearances, poor suction piping design, low flows, etc., can create thrust actions in a single-stage pump. If the thrust bearing is not large enough, a pump or seal failure results.

4.9 Hydraulic Axial Thrust in Multistage Pumps

Balancing hydraulic axial thrust in multistage pumps is more complex than in single-stage units. The axial imbalance is caused by varying pressure distributions on the front and back sides of the impeller. In most multistage pumps, there is no wearing ring mounted on the back shroud, and a pressure-force imbalance exists in the direction of the eye of the impeller. An additional axial load is caused by interstage leakage. Leakage will occur between adjacent stages because one stage is at a higher pressure than the other, as shown in Fig. 4.15. Pressure builds up on the back impeller shroud, causing an additional imbalance toward the eye of the impeller. This thrust imbalance caused by leakage happens at every stage, so the combined effect can be significant. The thrust imbalance will increase with an increase in bushing and wear ring clearance due to wear.

There are at least four different basic approaches to the axial thrust problem with a number of variations of each.

1. The balancing drum arrangement uses single suction impellers in series so that the flow moves progressively from one end of the pump to the other. This gives the simplest casing possible. Since all impellers face in the same direction, the resultant axial thrust is the sum of all unbalanced forces acting on each stage and is directed toward the suction. A balancing drum (Fig. 4.16)—basically a rotating piston with the characteristics of an pumped product-lubri-

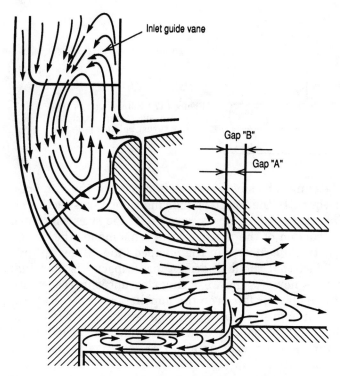

Figure 4.15 Leakage flows of multistage pumps.

cated radial bearing—is used to maintain axial balance. The balancing drum can influence the radial dynamic behavior of the pump rotor. It is a "soft" device in the axial direction. A major disadvantage of the design is that it requires considerable axial length. Balancing drum failures tend to result from other pump deficiencies, most commonly high levels of radial rotor vibration.

2. The balancing disk (Fig. 4.17) is basically a pumped product-lubri-

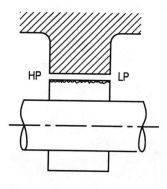

Figure 4.16 Balancing drum type of multistage pump axial hydraulic thrust balancing device.

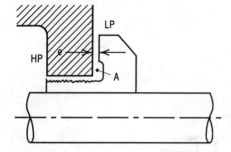

Figure 4.17 Balancing disk type of multistage pump axial hydraulic thrust balancing device.

cated thrust bearing. The small vertical gap controls the film pressure and consequently the counterbalancing rotor thrust action. If the gap becomes too small or closes completely, the faces will touch and cause destruction of the mating parts. This makes the design sensitive to particulate matter in the pumped fluid. Many failures are due to the fact that the disk design load capacity is marginal when the pump is new. As the wear ring clearances increase with normal use, the hydraulic forces on the impellers increase. Eventually this results in a greater force than the design thrust-reducing capacity of the disk. The balance disk, which relies on a small axial clearance to function properly, is a device sensitive to assembly variations and operational upsets. It reacts to all disturbances of the rotor, including mechanical dynamic imbalance, misalignment, vane passing frequencies, and subsynchronous rotor motion.

3. In the combination balancing drum and disk (Fig. 4.18), the drum is sized to accommodate about 85 percent of the thrust action of the pump, and the balance disk which is on the outboard end of the drum takes care of the remainder.

4. Opposing single suction impellers in equal groups permit the thrust of one impeller to neutralize that of another (Fig. 4.19). The opposed-impeller concept is frequently advertised as eliminating balancing devices. The opposite is true: there must be at least two balancing devices, not one. There is one interstage bushing with a

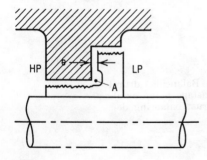

Figure 4.18 Combination balancing drum and disk type of multistage pump axial hydraulic balancing device.

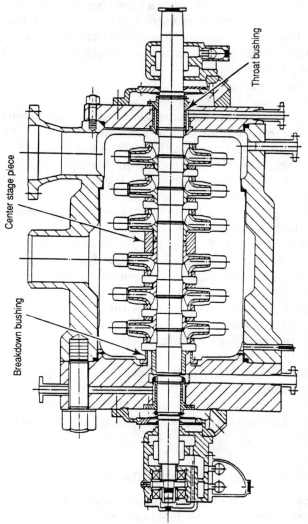

Figure 4.19   Axial hydraulic thrust balancing by use of opposed impellers.

considerable pressure differential across it, usually at the midspan of the rotor. Since this is the point of greatest rotor deflection, seizures are more frequent than in the straight-through design where the balancing drum is located near a bearing. A great number of combinations and arrangements of opposed stages are possible. The opposed-impeller design requires a more complicated casting for the casing with extra nozzles for entry and exit of the various stages.

Axial thrust in this method depends on the impeller arrangement and to what extent several factors are satisfied. These factors are

a. Stuffing box pressure

b. Interstage leakage

c. Casing design

A multitude of practical compromises must be made involving these factors. One compromise uses a breakdown throttling bushing to reduce the pressure acting on the stuffing box. Since the only flow across a throat bushing is the sealing leakage, a small amount, clearances used are usually about 0.030 to 0.035 in. The breakdown bushing requires only about one-third of the clearance because there is a higher rate of liquid flow across the breakdown bushing and out the balancing line back to the suction of the pump. In the pump shown in Fig. 4.19, diametral clearances are 0.012 to 0.014 in on the breakdown bushing (third-stage discharge pressure) and 0.030 in on the throat bushing (first-stage suction pressure).

Frequently, this bushing and the sleeve it runs against are not maintained properly. In these cases the stuffing box pressure increases, creating conditions for a premature failure of the mechanical seal and an increased thrust action of the pump. The importance of this breakdown bushing is not discussed in the pump maintenance manuals or in handbooks.

Because complete balance is never attained with opposed stages, an adequately sized thrust bearing is vital to take the residual axial thrust, which increases rapidly with wear in the running clearances.

4.10   Selection of Balancing Device Design

While all four types of multistage pump balancing devices can be made to operate satisfactorily, each has inherent shortcomings. The high pressure differential across the balancing devices causes considerable leakage of product to the pump's suction. The high pressure differential also means rapid wear of the balancing device's throttling surfaces and high maintenance. Disk-type balancing devices tend to fail more often in process service, while failures of the drum-type devices are rarer. In addition, when a balancing disk fails, the complete rotor is more likely to be destroyed than with the drum design. An adequately sized thrust bearing must be used with all four designs.

References

1. Dr. E. Makay and J. A. Barrett, "Changes in Hydraulic Component Geometries Greatly Increased Power Plant Availability and Reduced Maintenance Costs: Case Histories," *Proceedings of First International Pump Symposium,* Texas A&M University, Houston, May 1984.

Antifriction Bearings

5.1 Bearings--General

The bearings in a centrifugal pump must

- Allow the shaft to rotate with practically negligible friction
- Hold the rotating element in its proper position relative to the stationary parts of the pump, both radially and axially, so that rubbing cannot occur
- Be able to absorb the forces that are transmitted to them from the impeller
- Give trouble-free service for long periods of time

The types of bearings used in centrifugal pumps are categorized according to the direction of the forces they absorb, either radial or axial thrust. Centrifugal pump bearings may be antifriction bearings, ball or roller types, sleeve-type journal bearings, or hydrodynamic-type thrust bearings. Journal bearings carry radial loads only. Hydrodynamic thrust bearings can absorb loads in the axial direction only. Antifriction bearings are designed to handle a combination of both radial and axial thrust loads. Antifriction bearings are more popular in smaller, lower-speed pumps.

5.2 Antifriction Bearings

Antifriction bearings use balls or rollers instead of a hydrodynamic fluid film to support a shaft load with minimal wear and reduced friction (Fig. 5.1). Cleanliness, accuracy, and care are required when ball bearings are installed. The ball bearing is a piece of precision equip-

1: Oil formation

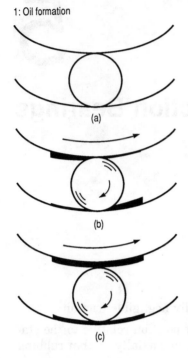

(a)

(b)

(c)

Figure 5.1 Theory of ball bearings: (*a*) Bearing static condition; (*b*) partial operating speed; (*c*) operating at full speed.

ment manufactured to extremely close tolerances. To obtain maximum service from a bearing, the shaft and housing must be machined to rigid tolerances also; e.g., locating shoulders must be at right angles to the shaft centerline so that the bearing will be squared with the shaft, and the housing bores must be in almost perfect alignment to ensure that the bearing will not be forced to operate in a twisted position. Maintenance of ball bearings is simple: protect the bearing from contaminants and moisture and provide proper lubrication.

5.3 Bearing Loads

Assuming that all of the above installation conditions are satisfied, the life of a ball bearing is dependent upon the load it must carry and the speed of operation. The loads on pump bearings are imposed by the radial and axial hydraulic forces acting on the impeller.

In any two-bearing system, one of the bearings must be fixed axially, while the other is free to slide. This arrangement allows the shaft to expand or contract without imposing axial loads on the bearings, and yet definitely locates one end of the shaft relative to the stationary parts of the pump. The outboard bearing of between bearings design pumps or the closest one to the coupling on a back pull out design is fixed axially. The inboard bearing is free to slide within the housing bore to accommodate the thermal expansion and contraction of the shaft, as shown in Fig. 5.2.

Since the outboard bearing is fixed in the housing, it must carry the axial thrust in addition to radial thrust. The axial thrust is considered to be acting along the centerline of the shaft, and therefore, is the same at the outboard bearing as it is at the impeller. The radial and axial loads combine to create a resultant angular load at the outboard bearing.

Radial thrust acting on the impeller creates radial loading on both bearings. The magnitude of the radial load at each bearing can be determined by the use of the following equations:

$$R_1 = \frac{P \times a}{s} \tag{5.1}$$

$$R_2 = \frac{P(a + s)}{s} \tag{5.2}$$

where R_1, R_2 = radial load at bearings 1 and 2, respectively, lb
P = radial thrust on impeller, lb
a = distance from centerline of impeller to centerline of inboard radial bearing, in
s = distance from centerline of inboard radial bearing to centerline of outboard bearing, in

Radial loads also come from other sources. The weight of the rotating assembly (shaft, sleeve, and impeller) gives one load. Imbalance and

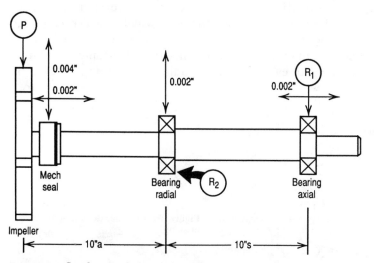

Figure 5.2 Loads on radial and thrust bearings.

shaft external misalignment give another. The weight of the overhung coupling also gives a bearing load, as discussed under hydraulic radial forces. Pump designs should limit the shaft deflections at the stuffing box face to under 0.002 in at the worst conditions. For single-stage horizontal pumps, this will be with the maximum impeller diameter and at "shutoff" conditions, i.e., closed discharge. For larger double suction pumps, this load might well occur at the far end of the performance curve. Attention paid to cutwater clearances, Gap "B," and "overfiling" of vanes can reduce some of the hydraulic loads. The details of this subject area are discussed in Chap. 8.

5.4 Types of Ball Bearings

A ball bearing normally consists of two hardened steel rings, with several hardened balls utilizing a separator to space the rolling elements and reduce friction, as shown in Fig. 5.3. Many types of ball bearings are available and are classified according to the type of loading: radial, thrust, and combined loading. Sizes and classes of precision of bearings are governed by the Anti-Friction Bearing Manufacturing Association (AFBMA) and by the Annular Bearing Engineers Committee (ABEC). There are five ABEC classes—1, 3, 5, 7, and 9. Class 1 is standard, and class 9 is high-precision. Pump bearings generally are class 3, loose fit. ABEC class 9 is factory-order only and has no longer bearing life or higher speed rating than ABEC 1. Three types of bearings are generally used in centrifugal pumps.

5.4.1 Conrad type

The Conrad type (identified also by its design features as the deep-groove or nonloading groove type) is the most widely used (Fig. 5.4). A general-purpose bearing, it is used on electric motors or wherever slight axial movement of the shaft is permissible. Its deep-groove rings permit this bearing to carry not only radial loading, for which it is

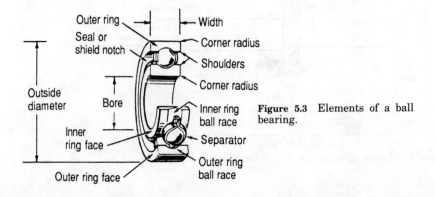

Figure 5.3 Elements of a ball bearing.

Single-row
deep-groove
super-conrad

Figure 5.4   Conrad bearing
design.

primarily designed, but also about 75 percent of that amount of thrust
load in either direction, in combination with the radial load. The API
610 standard, *Centrifugal Pumps for General Refinery Use,* 7th ed.,
requires that single-row or double-row radial bearings be Conrad type
with class 3 or loose fit. This permits enough flexibility to let the shaft
correct for any misalignment between the housing and the shaft.

5.4.2   Maximum-capacity type

Another radial bearing type is the maximum capacity or filling-slot
type. It is provided with a filling notch that extends through the ring
or ring shoulders to the ring way, and it permits a larger number of
balls to be placed between the rings than can be done in the same size
Conrad-type bearing. The supposed advantage of filling-slot bearings
(larger load-carrying capability) has vanished with the availability of
better steels and lubricants. Filling slots often are not precision-ma-
chined and can enter the ball contact area of the rings. This will result
in early bearing failure. Several bearing manufacturers consider a
filling-slot bearing to be unreliable and recommend against its use.
The 1981 and later editions of API 610 prohibit their use, although
they are still permitted by ANSI specifications.

5.4.3   Angular contact type

This design allows the carrying of high radial loads in combination
with thrust loads from 150 to 300 percent of the imposed radial load.
Unlike the Conrad design, in this design the contact angle is not perpendicular to the bearing axis. Several different angles are available,
offering a variety of radial and thrust loadings. With this bearing
design, filling slots are not used. Thrust loads may be imposed from one
direction only.

Duplex-type bearings are identical angular contact bearings placed
side by side. The contacting surfaces must be ground to generate a
specified preload (Fig. 5.5). This special grinding allows the two bearings to share loads equally. Without it, one bearing in the pair would
be overloaded and the other underloaded.

API pump specifications require duplex 40° contact angle thrust
bearings mounted back to back with a light [100-lb or 45-kilogram

Angular-contact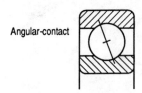

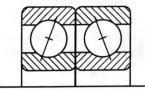

Figure 5.5 Angular contact
bearing design.

(kg)] preload as the best choice as shown in Fig. 5.6. Although the
requirement of a 7000 series, 40° contact angle, light preload bearing
should be a fairly tight specification, it is not. First, there are three
7000 series bearing designs: light, medium, and heavy. There is about
a 50 percent change in capacity between each of the designs. Second,
some manufacturers use more than one contact angle. The contact
angle is the source of considerable confusion, as shown in Fig. 5.7.
There are no standard designations to identify the 40° angle. Not all
ball bearing companies manufacture the 40° angular contact bearing.
If the local bearing supply house is not on its toes, pump repairs may
be made with nonspecification bearings. The 40° contact angle gives an
18 to 40 percent increase over the 30° angle depending on the bearing
size. This capacity change is a very important piece of information to
be communicated by a confusion of suffixes. The numerical code used
in bearing identification is mostly standard among the various manufacturers. However, the alphabetical prefixes and suffixes are not.
Care should be exercised when you are identifying bearings from
codes, for the purpose of interchanging bearings, that the meaning of
all numbers and letters is ascertained so that an exact substitution can
be made. Most manufacturers supply cross-reference tables for identifying equivalent bearings. A very good source is the bearing manual of
the AFBMA.

Duplex
bearings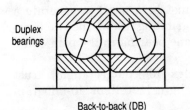

Figure 5.6 Back-to-back duplex
mounting of angular contact
bearings.

Back-to-back (DB)

Contact angles for 7000 series duplex bearings					
Bearing manufacturer	Contact angle				
	20°	25°	30°	35°	40°
A	X			X	X
B			X		
C		X		X	
D					X

Figure 5.7 Contact angles of various bearing manufacturers.

5.4.4 Mountings

There are five different types of duplex angular contact bearing mountings, although only two are commonly used. The rigidity of the shaft and bearing assembly depends in part on the moment arm between ball contact angles of the duplex bearings. Within reasonable limits, the longer the moment arm, the greater its resistance to misalignment.

API 610 calls for the DB, or back-to-back, mounting of angular contact bearings. They are placed so that the stamped backs of the outer rings are together. In this position, the ball contact angles diverge outwardly, away from the bearing axis. With DB bearings, the space between the diverging contact angles is long, as shown in Fig. 5.8. Shaft rigidity is correspondingly increased, and resistance to misalignment is increased.

DF bearings are intended only for face-to-face mounting. They are

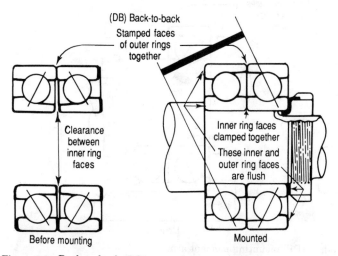

Figure 5.8 Back-to-back (DB) mounting moment arm.

placed so that the faces (or low shoulders) of the outer rings are to-gether. Ball contact angles converge inwardly, toward the bearing axis. With DF bearings, the space between the converging contact angles is short. The bearing shaft rigidity is relatively low. However, this arrangement permits a greater degree of shaft misalignment than other mounting methods do (Fig. 5.9). Some older multistage pumps use this mounting arrangement.

5.4.5 Special designs

Centrifugal force in the unloaded thrust bearing causes the balls of that bearing to move out of their intended track and operate in a skewed axis. The balls begin to slide rather than roll during rotation. The increased friction causes reduced viscosity of the oil film and ac-celerated wear of the raceways, leading to early failure. Some recent work in bearing selection indicates that adherence to the API specifi-cation of 40° bearings mounted back to back may not be the optimum selection. For single-stage pumps in which the thrust action is steady and in one direction at all flows, the use of a 40° angular contact bearing to absorb the primary thrust and a 15° angular contact bearing for any reverse thrust has resulted in extended life for pump bearings. The 15° bearing decreases the tendency for ball sliding and increased friction. The bearings also have machined bronze retainers to further reduce internal friction. The bearings are packaged in pairs and are marketed so as to mount them to accommodate the primary thrust load (Fig. 5.10).

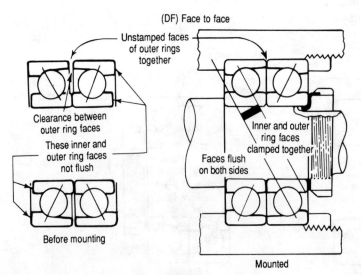

Figure 5.9 Face-to-face (DF) mounting moment arm.

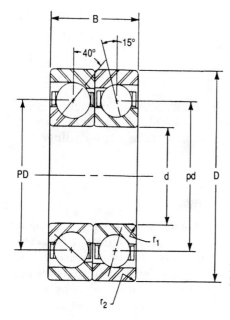

Figure 5.10 Special design duplex angular contact bearing.

This arrangement is better in some but not all applications. Recommendations for its use are summarized as follows:

1. *Pump:*   overhung single suction pump and low speeds, 1750 cycles/min or below, with any load conditions. Bearing style: 40° duplex, DB mounting, or 40° and 15°.

2. *Pump:*   overhung single suction pump and high speeds, in excess of 1750 cycles/min, with large thrust loads coupled with radial load. Bearing style: 40° and 15°, DB mounting.

3. *Pump:*   double suction between bearings and high speeds, in excess of 1750 cycles/min, mostly radial loads with low thrust loads. Bearing style: duplex 40°, DB mounting.

5.4.6 Double-row type

Essentially, the double-row bearing is an integral duplex pair of angular contact bearings with built-in preload (Fig. 5.11). It resists radial loads, thrust loads, or combined loads from any direction. Two basic types are available, corresponding to the face-to-face and back-to-back mounting of conventional duplex bearings. Avoid the use of two double-row bearings on the same shaft as this makes for too rigid a mounting. The bearings on each end of the shaft will tend to impose loads on each other.

Figure 5.11   Double-row bearing.

While double-row bearings can be constructed to be angular contact designs, they require a filling slot for assembly of at least one row. API 610 prohibits the use of the design because of the greater vulnerability to failure in thrust applications.

5.5   Bearing Misalignment Capability

The ability to tolerate misalignment between the bearing housing and shaft is dictated by the ball and ring geometry. Figure 5.12 is a chart showing the relative capabilities of three different bearing types. Knowledge of these relative capacities can equip a maintenance engineer or supervisor to consider substitutions to solve many bearing problems. Note that for the Conrad bearing the rated radial capacity and speed limit are taken as unity for comparison purposes.

The angular contact bearing can carry almost double the Conrad's radial rating in thrust. The Conrad can carry only 75 percent of its radial rating in thrust, and the self-aligning ball bearing can carry only 20 percent. This means that the thrust load on a self-aligning ball bearing is prohibited. Note the angular misalignment capability of the various bearings. The Conrad can stand 15′; the self-aligning ball can stand 16 times as much, or 4°. Values in this chart are for comparison purposes only. Actual catalog values for load ratings, limiting speed, etc., should be used.

5.6   Other Bearing Problems

A number of problems are associated with the use of antifriction bearings that affect the reliability of a pump design.

Type	Average relative ratings			
	Radial	Thrust	Limiting speed	Misalignment
Conrad type	1.00	0.75	1.00	±0 deg 15′
Angular contact 40°	1.00	1.90	1.00	±0 deg 2′
Self-aligning	0.70	0.20	1.00	±4 deg

Figure 5.12   Capacity chart for various bearing designs.

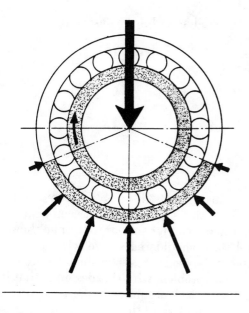

Figure 5.13 Load zones and retainers of a ball bearing—360° for inner ring and 150° for outer ring.

5.6.1 Retainers

A retainer ring or cage is used to make all the balls of a bearing go through the load zone (Fig. 5.13). The most common retainer material is low-carbon steel (1010 analysis), attached by fingers, rivets, or spot welding. Riveted or spot-welded steel strip retainers are more subject to fatigue failures. When a bearing ring is misaligned, the balls are driven up against the ring shoulder, the top ball to the left and the bottom ball to the right. The center balls on each side, at this particular point, tend to stay in the center of the ring because balls in this position relative to the misalignment are not thrust-loaded. The net effect of this action is to flex the ring in plane bending. As the inner ring turns, a cyclic retainer bending stress occurs. The load on the retainer pocket is also cyclic. At the high-thrust positions, the retainer exerts the maximum force in maintaining the ball space. Since the retainer and ball are in rubbing contact, the thermal load is at its highest on one side (no thrust load point). In this manner, the retainer is subjected to both a flexing and thermal cyclic load that can lead to fatigue cracking at retainer stress points such as forming notches and rivet holes. Due to the *rubbing* contact between the retainer and balls, the lubrication requirements here are more critical than for the *rolling* contact between the balls and rings. Shock loading of the bearings also causes retainer failure at the pockets.

Some manufacturers use pressed brass, machined bronze, machined phenolic, and molded plastics in an effort to reduce the heat generation.

Bearing separator types are ranked according to desirability:

1. Phenolic
2. Machined bronze
3. Pressed brass strips
4. Pressed steel strips
5. Riveted steel strips

5.6.2 Bearing carriers

In order to use a larger radial bearing and still be able to remove the impeller or mechanical seal from the shaft, some between-bearing pump manufacturers utilize a bearing carrier similar to the one shown in Fig. 5.14. The carrier itself is a shouldered sleeve with a small clearance fit onto the shaft. The radial bearing is then shrunk onto the outside diameter of the carrier. The problem with this design is that if the bearing begins to heat up, due to lack of lubrication or some other reason, the carrier also heats up, expanding until it loses its fit. At this point, even though the bearing has not failed, the carrier may be free to spin on the shaft, which will eventually cause the shaft to bend or fail.

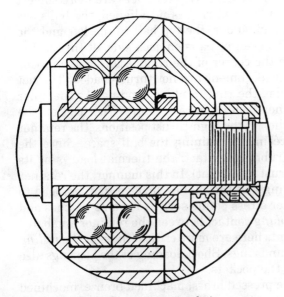

Figure 5.14 Typical bearing carrier design.

5.6.3 Snap rings

Snap rings are flat, split, washerlike devices used by some manufacturers to axially position components, e.g., ball bearings and seal sleeves on shafts (Fig. 5.15). As with bearing carriers, there are two major problems associated with the use of snap rings. First, removal requires the use of a tool that is not normally found in the pump machinist's toolbox. When used in a shaft, snap rings must be positioned in a groove cut in the shaft. The addition of a radial groove in the shaft effectively reduces the diameter and may weaken the shaft. When used in a bearing mounting, the rings permit considerable end float of the bearing. API 610 specification does not permit snap-ring-mounted bearings.

5.7 Bearing Arrangements

Different arrangements of antifriction bearings can handle various loadings imposed on the pump. The design of the pump is crucial to the bearing arrangements.

5.7.1 Horizontal pumps

Overhung impeller pumps usually employ ball bearings only. In a typical bearing housing arrangement (Fig. 5.2), the radial ball bearing is located adjacent to the impeller or inboard position. It is arranged to take only radial loads. The thrust bearing is located closest to the

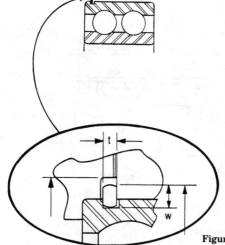

Figure 5.15 Snap ring mounting.

coupling and usually consists of a duplex pair of angular contact bearings. The bearings are mounted in a back-to-back arrangement such that axial thrust load can be carried in either direction. This duplex bearing pair carries both the unbalanced axial thrust loading and the radial load.

In between-bearing pumps, the ball radial bearing and the ball thrust bearing combination have individual bearing housings (Fig. 5.16). The radial bearing is normally located at the coupling end of the pump. The ball thrust bearing is located at the outboard pump end. The thrust bearing must be secured axially on the shaft to transmit the axial thrust load to the bearing housing through the bearing. The bearing is usually located against a shoulder on the shaft and locked in place by a bearing nut. This means that the shaft diameter under the thrust bearing is less than the shaft diameter under the radial bearing. Thus, by mounting the radial bearing on the inboard, or coupling, end of the pump shaft, a larger shaft diameter is available to transmit pump torque from the coupling to the impeller. Whereas the thrust bearing is locked axially in the thrust bearing housing, the radial bearing is axially loose in its housing to allow for axial thermal growth.

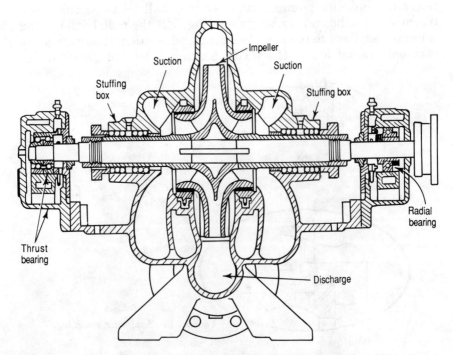

Figure 5.16 Typical pump bearing arrangements—between bearing pump.

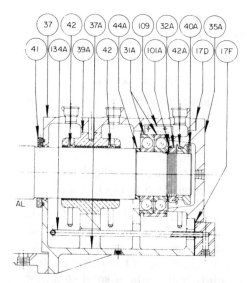

Figure 5.17 Combination sleeve radial and ball thrust bearings.

A popular combination for between-bearing double suction pumps consists of journal-type radial bearings and a ball thrust bearing (Fig. 5.17). In such an arrangement, all radial pump loads are handled by the journal radial bearing. The ball thrust is mounted in the thrust bearing housing such that only axial loads are carried by the thrust bearing. The housing around the ball thrust bearing is radially loose. A metallic strap is employed on the outer rings of the thrust bearing which locks into the bearing housing to prevent rotation of the outer rings. This bearing arrangement is useful in higher-horsepower and higher-speed applications where ball radial bearings would be impractical due to speed, load, and lubrication limitations. Because the ball thrust bearing is located on the outboard end of the shaft, the shaft diameter under the ball thrust bearing can be relatively small since no torque is transmitted to this end of the shaft.

5.7.2 Vertical pumps

Most vertical pumps differ from horizontal pumps in that the entire axial thrust, consisting of axial hydraulic forces as well as the static weight of the pump *and* the driver rotor, is supported by the driver thrust bearing. Therefore, the sizing of that bearing becomes a joint effort by the pump manufacturer, the driver manufacturer, and the end user.

Many end users require that this bearing be rated to handle at least twice the maximum thrust load, up or down, developed by the pump when in the worn condition with two times the new internal clearances. This basic requirement was added after numerous problems were found in the field that are a direct result of the following:

1. The calculation of the pump thrust is not highly accurate.
2. The pump thrust increases as internal clearances increase.
3. The thrust load varies with the vertical position of the impeller(s) with the casing(s).
4. The thrust load varies with flow; in some cases, it may even reverse direction.
5. A reasonable margin should be provided between the driver thrust bearing rating and the maximum calculated pump thrust.

Motors for vertical pumps are available as solid-shaft or hollow-shaft units. On hollow-shaft units, the pump shaft extends upward through the motor shaft and is supported at the top of the motor shaft. Clearance is provided between the outside diameter of the pump shaft and the bore of the motor shaft. On solid-shaft units, a solid coupling is provided by the pump manufacturer to provide rigid attachment between the pump shaft and the motor shaft extension. Thus, the pump shaft is retained radially by the lower motor bearing. This is considered to be the better arrangement for the majority of vertical pumps since the shaft runout will be less and thus the seal or packing life will be longer. Further, larger-diameter pump shafts can be coupled to solid-shaft motors than can pass through the bore of hollow-shaft motors, and this also provides increased shaft rigidity.

Because of field problems, the thrust bearing should be mounted in the driver top bearing housing, farthest from the solid coupling and the pump (Fig. 5.18). In the event that a thrust bearing fails, any subsequent drop in the driver or pump shafts could result in a mechanical seal failure that could release hydrocarbons to the atmosphere and be ignited by a hot bearing.

In many vertical pumps, it is common to position the shaft with pump-lubricated sleeve bearings (Fig. 5.19). In smaller pumps, especially small multistage vertical pumps, these bearings are normally of simple sleeve design and are arranged to ensure a flow of product lubrication through the bearing clearances. On water vertical pumps, cutlass rubber bearings, with large flutes for passage of lubrication water, are commonly employed.

5.8 Ball Bearing Fits

Unfortunately, many pump manufacturers do not indicate the proper bearing fits for shaft and housings to guide shop repairs. The original dimensions of both the housing and the shaft will change from time to

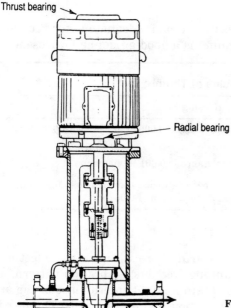

Thrust bearing

Radial bearing

Figure 5.18 Vertical pump motor—solid shaft.

time due to oxidation, fretting, damage from locked bearings, and other causes. Every bearing handbook has tables to aid you in selecting fits. The vibrational effect of looseness on the bearing fits is different for the housing and the shaft.

5.8.1 Housing fits

Ball bearing fits in the bearing housing are of necessity slightly loose for assembly. If this looseness becomes excessive, vibration at rotational speed and multiple frequencies will result. Do not install bearings with outside diameter (OD) outside the given tolerance band since

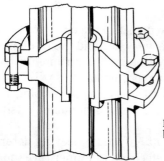

Figure 5.19 Vertical pumps—bushings.

this might result in either excessive or inadequate outer race looseness. The following rule of thumb is a good guideline for looseness.

Rules of Thumb:
Housing Fits
1. Bearing OD-to-housing clearance is about 0.00075 in loose with 0.0015 in maximum.
2. Bearing housing out-of-round tolerance is 0.001 in maximum.
3. Bearing housing shoulder tolerance for a thrust bearing is 0 to 0.0005 in per inch of diameter off square up to a maximum of 0.002 in.

5.8.2 Shaft fit

A loose fit of the shaft to the bearing bore will give the effect of an eccentric shaft, at 1 times running frequency vibration pattern. The objective of the shaft fit is to obtain a slight interference of the antifriction bearing inner ring when mounted on the shaft. The bearing bore should be measured to verify the inner race bore dimensions. Do not install bearings with an inside diameter (ID) outside the given tolerance band, since this might result in either excessive or inadequate shaft tightness. The rules of thumb below give good guidelines.

Rules of Thumb:
Shaft Fits
1. Fit of bearing inner race bore to shaft is 0.0005 in tight for small sizes and 0.00075 in tight for large sizes.
2. Shaft shoulder tolerance for a thrust bearing is 0 to 0.0005 in per inch of diameter off square up to a maximum of 0.001 in.

5.9 Detection of Antifriction Bearing Defects

Antifriction bearing defects are difficult to detect in the early stages of a failure because the resulting vibration is very low and the frequency is very high. If monitoring is performed with simple instrumentation, these low levels will not be detected and unexpected failures will occur. The vibration frequencies transmit well to the bearing housing because the bearings are stiff. Detection of defects is best done by using accelerometers or shock pulse meters.

Here are some guidelines that can help to evaluate bearing deterioration. For example, a ball passes over defects on the inner race more often than on the outer, because the linear distance around the diameter is shorter. There are four dimensions of a ball bearing that can be used to establish some feel for the condition of that bearing:

1. A defect on outer race (ball pass frequency outer) occurs at about 40 percent of the number of balls times running speed.

2. A defect on inner race (ball pass frequency inner) causes a frequency of about 60 percent of the number of balls multiplied by running speed.

3. Ball defects (ball spin frequency) vary with lubrication, temperature, and other factors.

4. Fundamental train frequency (retainer defect) occurs at lower than running-speed values.

A simple check for verification of poor bearing condition is made by shutting off the pump and observing that the high bearing frequency remains as the pump speed reduces. This high-frequency signal will normally remain until the pump stops. The frequency indication is normally from 5 to 50 times the running speed of the machine.

5.10   Lubrication of Antifriction Bearings

Antifriction pump bearings can be either grease- or oil-lubricated. Failure from lack of effective lubrication, in either type or quantity, constitutes a major source of bearing difficulties. The primary purpose of oil, or the oil constituent of grease, is to establish an elastohydrodynamic film between the bearing's moving parts, as shown in Fig. 5.1. This oil film is the result of a wedging action of the oil between the roller elements and raceways. The formation of this film is, to a major degree, a function of the bearing operating speed and, to a lesser degree, the magnitude of the applied load. Lubrication for ball- or roller-type bearings can be developed in three ways:

1. Full elastohydrodynamic

2. Complete lack of separating oil film

3. Mixture of the above conditions or boundary-type lubrication

While the surfaces of bearings are highly finished, there are peaks and valleys. Use of correct-viscosity lubricant ensures development of a full oil film between rotating parts. In boundary-type lubrication, a metal-to-metal contact condition occurs and friction wear develops. If the bearings are operated with the correct viscosity lubricant for the speeds and loads involved, a full elastohydrodynamic film will develop between the rotating parts, as shown in Fig. 5.1. Under these conditions the oil film formation has enough thickness to completely separate the unevenness on the bearing surfaces.

Since there is no metal-to-metal contact with full film lubrication, there is no wear on the bearing parts. The only time that metal-to-metal contact occurs is on start-up or when the bearing is brought to rest.

A lubricant that has a viscosity too low for the operating loads and speeds permits penetration of the oil film, which results in moving surfaces making direct contact. In boundary-type lubrication, this metal-to-metal friction causes wear of the surfaces to increase rapidly, as the film is frequently ruptured. Viscosity requirements for both ball- and roller-type bearings are equated in terms of DN value (a factor used to compare the speed effects of different-sized bearings). This value is obtained by multiplying the bearing bore size in millimeters by the actual rotating speeds in revolutions per minute. It is important to remember, however, that the advantages of proper viscosity can be offset at high speeds if too much lubricant is placed in the bearing and housing cavity. At high speeds, excessive amounts of lubricant will create a churning action of the oil and increase the friction and operating temperatures of the bearing.

Other functions of a lubricant are to assist in protecting highly finished bearing surfaces from corrosion and, in the case of grease, to aid in the exclusion of foreign contaminants from the bearing chamber through periodic regreasing. Increased operating temperature reduces oil viscosity, film thickness, and accelerated deterioration of the lubricant. Petroleum-based lubricants operated beyond the 200°F range will have their life reduced by 50 percent for every 18°F temperature rise above this level. The lubricants tend to carbonize or harden within the bearing cavity. Deterioration is caused by the initial evaporation of the higher volatile component parts of the oil, or the oil mixed in the grease.

5.10.1 Grease

Grease lubrication is normally limited to small, low-horsepower, noncritical pumps operating at relatively low speeds and temperatures (Fig. 5.20). The grease can be either located in the bearing housing surrounding the bearing or packed in the bearing and then sealed.

5.10.2 Oil-flooded

A more common lubrication system for centrifugal pumps is the oil flood (Fig. 5.21). In such an arrangement, the bearing housing provides a sump or an oil reservoir. This sump maintains a level of oil at or near

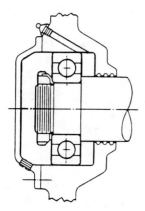

Figure 5.20 Grease lubricated pump bearing.

the centerline of the lowest ball of the bearings. This oil level is usually kept constant by means of a constant-level oiler. There are two problems with this type of lubrication. First, if the level is too high, frothing and foaming may occur, generating heat within the reservoir. Second, there is a very small range between the "proper level" and the bottom of the balls below which all further lubrication stops.

5.10.3 Ring-oiled

Ring-oiled systems are often used for lubrication of antifriction bearings for larger horizontal pumps where, because of speed or loads, a simple flood system is not adequate. Located axially between but adjacent to the bearings, rings of a diameter larger than the shaft ride on top of the shaft and dip into the oil reservoir below (Fig. 5.22). The oil

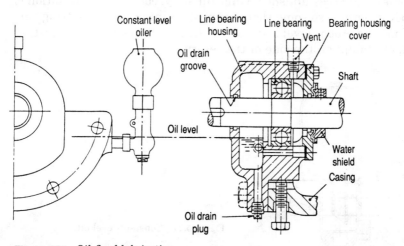

Figure 5.21 Oil-flood lubrication.

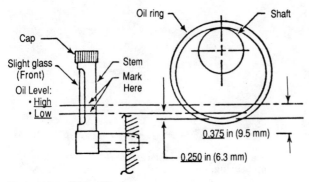

Figure 5.22 Ring oiling.

level in the reservoir must be maintained so that at least ¼ to ⅜ in of the ring bore is immersed. The rotation of the shaft causes the rings to rotate and carry oil from the reservoir up to the shaft. The oil is then thrown from the shaft by oil flingers, located adjacent to each oil ring, directly into each bearing to ensure complete lubrication. As the oil is circulated through the bearings, it is returned to the oil sump.

The constant-level oiler (Fig. 5.23) is a device used with both the oil-flooded and ring-oiled lubrication systems. It acts as a small reservoir for extra oil while maintaining a predetermined level in the bearing housing. The constant-level oiler uses the principle of a liquid seal to maintain the oil in the bearing reservoir constant. When, due to consumption, the oil level in the bearing recedes, the liquid seal on the spout is temporarily broken. This lets air from the air intake vents enter the oiler reservoir, thus releasing oil until the liquid seal and the proper level are reestablished. Unfortunately, each oiler installation is slightly different, so some thought must go into properly setting this position. The proper level is usually indicated (by a nameplate, casting mark, or stamp) on the side of the reservoir.

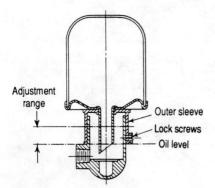

Figure 5.23 Constant-level oiler.

5.10.4   Oil mist lubrication

The basic concept of the oil mist lubrication system is dispersion of an oil aerosol conveyed into the bearing housing. The equipment necessary for such a system is shown in Fig. 5.24. There are two types of inlet fittings or reclassifiers of the oil particles. They differ in the degree of coalescence from essentially none, for the pure mist, to partial, for the purge mist, as shown in Figs. 5.25 and 5.26.

5.10.4.1   Pure mist   In the pure mist lubrication (Fig. 5.25), an oil-air mist, under pressure, is fed directly into the bearing housing. There is no reservoir of oil in the housing, and oil rings are not employed. The pressurized mist flows through the bearings. The moving components of the ball bearings produce internal turbulence, causing impingement and collection of oil on the surfaces of the ball bearing. Vents are located on the backside of each bearing to ensure that the mist travels through the bearings. A drain is located in the bottom of the bearing housing to prevent the buildup of condensed oil, as shown in Fig. 5.25. The advantage of pure mist is that it creates an uncontaminated environment in which the bearings may operate, and it protects them from adverse environmental conditions while, effectively eliminating heat buildup.

The oil mist system will follow the path of least resistance. The back-to-back mounting of the angular contact thrust bearing will have the most windage, so most of the flow of a single inlet fitting will go through the radial bearing which has less windage and hence less resistance to flow. As a result, heavily loaded bearings may require two spray inlets. The mist must flow from the inlet fitting through the bearing, then to the vent. The flow should be in the same direction as the thrust for duplex angular contact bearings, as shown in Fig. 5.27. The positions of the vent and the center spray can be interchanged. Vent area should be about twice the reclassifier bore area. This will create a slight back pressure in the bearing housing to keep dirt out. All vents should carry about an equal portion of air in multivent installations. Different-size reclassifier orifices are needed according to bearing size.

During the mounting process, bearings must be heated to about 250°F to go on the interference fits of the shaft. Most of the oil on the bearing will flow off. To replace this oil, two approaches may be taken. First, the pump bearing should be reoiled by filling the bearing housing with oil up to the shaft level. The shaft should be turned 3 or 4 revolutions (r) so that the bearing is coated. Drain the oil out of the housing. Second, the oil mist system should be connected and operated 8 to 12 hours (h) to "plate out" an oil film on the bearing.

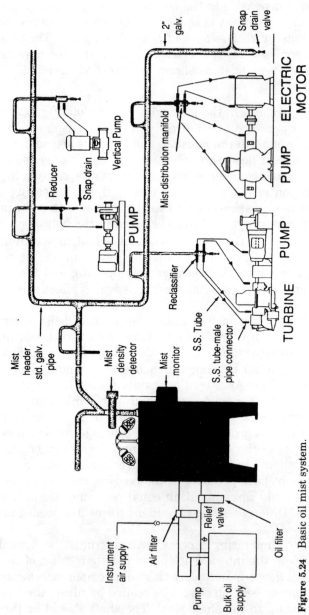

Figure 5.24 Basic oil mist system.

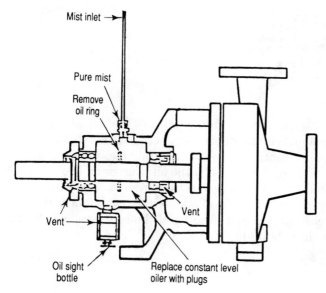

Figure 5.25 Oil mist lubrication—pure mist.

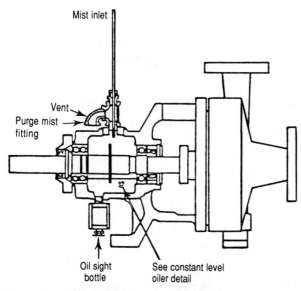

Figure 5.26 Oil mist lubrication—purge mist.

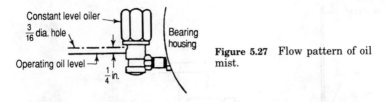

Figure 5.27 Flow pattern of oil mist.

5.10.4.2 Purge mist.   Another version of oil mist is called *purge mist* (Fig. 5.26). This system is employed in conjunction with a conventional oil-flooded or ring-oiled lubrication system. It combines the advantages of the positive oil circulation created by the oil rings or oil flood system with the pressurized uncontaminated oil mist system. When this combination is employed, a constant-level oiler with an overflow feature is employed to prevent buildup and flooding of the bearings, which can result in excessive heat buildup (Fig. 5.28).

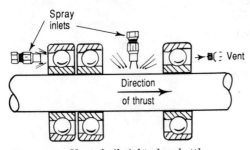

Figure 5.28 Vented oil sight glass bottle.

Hydrodynamic Bearings

6.1 Journal or Sleeve-Type Bearings

A sleeve or journal bearing provides radial support and positioning of the shaft while permitting it to rotate. It is composed of two principal parts: the rotating, cylindrical journal which transfers the shaft load and the stationary bearing and surrounding shell which supports the load (Fig. 6.1).

Journal bearings function because of the internal resistance of the lubricating oil to shear (viscosity). When the shaft rotates faster than

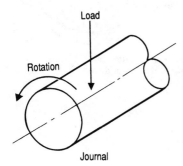

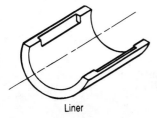

Figure 6.1 Basics of sleeve (radial) bearing.

about 25 surface feet per minute (ft/min), this shearing action creates an oil pressure wedge that lifts and pushes the shaft to a stable position in the bearing clearance. A journal bearing is dependent on the lubricant's ability to adhere to the surfaces of the journal and bearing. With the shaft at rest (Fig. 6.2), practically all of the lubricant has been squeezed out from under the shaft. When the shaft begins rotating, with the clearance space filled with oil, there is a tendency for the shaft to climb up the bearing sidewall due to friction (Fig. 6.3). Further rotation causes the oil to be drawn into the clearance space, lifting the shaft. The shearing action of the oil causes pressure to be developed in the wedge-shaped clearance space, and the shaft is forced over to the opposite side of the bearing (Fig. 6.4). In this position, the oil film pressure profile shown in Fig. 6.5 supports the load carried by the shaft.

Oil is used to lubricate and cool the bearing. Oil flows in journal

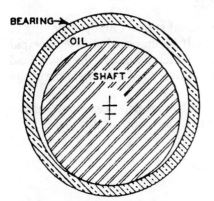

Figure 6.2 Shaft and oil film— at rest condition.

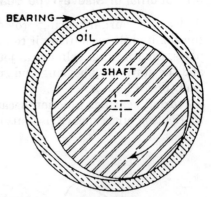

Figure 6.3 Shaft and oil film during start-up condition.

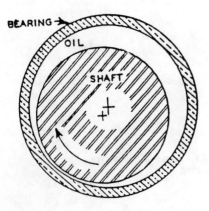

Figure 6.4 Shaft and oil film at running condition.

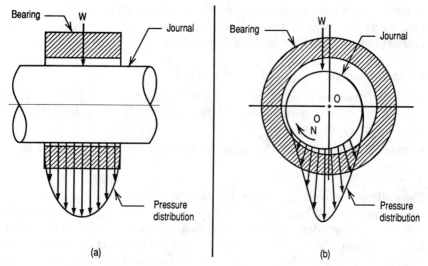

Figure 6.5   Oil film pressure profile. (*a*) Axial; (*b*) circumferential.

bearings are radially inward to a distribution groove where it contacts the rotating shaft and is drawn into the clearance area of the bearing. Normally there is an inlet for the lower and upper halves of the bearing. Almost all of the oil is expelled out the ends of the bearing. The cooling oil passes over the bearing system parts subjected to the least amount of pressure, absorbs the heat, and carries it back to the oil reservoir after being expelled out the distribution grooves. The oil that lubricates forms the film between the journal and the bearing. This oil film is subjected to great pressure. As it is supplied at little or no pressure, the force that maintains the film originates in the bearing itself and is generated by the pumping action of the journal revolving in the bearing. Minimum oil films will often be in the 0.0002 to 0.0004 inch range; therefore, the elimination of all abrasive particles larger than this clearance (about 5 to 10 μ) is essential to minimize wear.

Figure 6.5 shows the pressure distribution in the oil film of a journal bearing running under load. Note that the shaft center does not coincide with the bearing center, but lies below and to the left of the vertical and horizontal centerlines. The distance between the centers is known as the bearing's *eccentricity*.

6.2 Bearing Surfaces

The bearing surface is the part of a journal bearing that performs the basic antifriction function and thus is of primary importance. Sir Isaac Babbitt developed a bearing metal in 1839. Almost the same composition is commonly used today in larger pumps: 89 percent tin, 8 percent antimony, and 3 percent copper. This alloy is an almost ideal journal bearing surface because it provides

- The slipperiness required to overcome friction
- The softness required to permit a reasonable amount of foreign particles to embed themselves and to limit the damage to the more expensive shaft
- Deformability so that the shaft and bearing will conform to each other

The ability of a bearing material to deform without causing a failure is very important. When loads are applied, the bearings and journals deflect, causing edge loading of the bearing surface. Periods of high load are generally of short duration. Tin-based babbitt is able to wipe locally, relieving any high spots or areas of contact, and then deflect so as to reestablish fluid-film conditions. The bearing then frequently settles down for an extended period of good operation. Bearing materials that have good deformability or flow also have low fatigue strength. Since a high degree of deformability is essential, some compromises must be made.

Babbitt can be divided into two categories, conventional and microbabbitt or thin babbitt. Conventional babbitt bearings have substantially heavier thickness of babbitt and provide excellent conformability, embedability, surface action, and corrosion resistance, but poor fatigue strength. Thin-shell babbitt bearings offer excellent fatigue strength, good surface action, and corrosion resistance, but poorer conformability and embedability than conventional babbitt. The bearing life increases rapidly as the babbitt thickness decreases below 14 mils, as shown in Fig. 6.6.

Babbitt loses strength rapidly with increasing temperature. Its strength at 200°F is one-half that at 100°F, so the maximum babbitt temperature must be kept below about 225°F. Drain oil temperature should be below 180°F.

6.3 Bearing Crush

Thin-shell or insert-type bearings must be fitted with crush built into them. The term *crush* means that the split faces of the half bearings are made to project a certain amount beyond the split faces of the bearing housing, thus ensuring a tight fit of the bearings in their housing when its bolts or screws are properly tightened (Fig. 6.7).

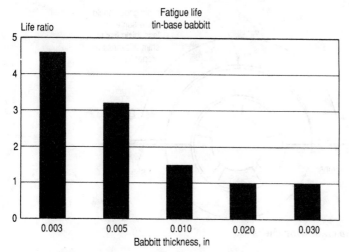

Figure 6.6   Bearing life as affected by babbitt thickness.

As a general rule, a minimum diametral crush of 0.001 to 0.002 in is satisfactory for most pump applications. This value, however, may be varied to satisfy certain assembly conditions influenced by housing material and design.

6.4 Bearing Clamp

Thick-shell bearings and the holder for thin-shell bearings must also be provided with a clamp or interference fit between the outer bearing *shell* and the bearing housing. Depending on the size and design, the

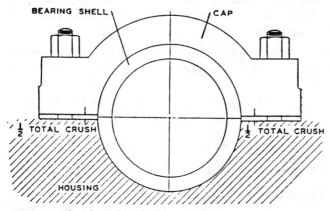

Figure 6.7   *Crush* on thin-shell bearings.

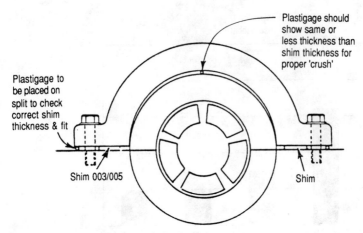

Figure 6.8 *Clamp* on bearing shells.

clamp should be between 0.0005 and 0.001 in. The procedure for measuring clamp by using metal or plastic gaging material is shown in Fig. 6.8.

1. Install bearing in the lower housing.

2. Add equal shim stock to either side of the housing split line. Make sure that the shims straddle the bolt holes. The thickness of the shims should be equal to or thinner than the diameter of the plastic or metal gaging material to be used.

3. Lay a short piece of measured gaging material across the top of the bearing shell. For long bearings and cylindrical shells, place a short piece at both ends of the bearing. The thickness of the gaging material should always be equal to or thicker than (no more than twice) the split-line shims.

4. Install the bearing cover and torque the bolts to the proper values. Remove the housing cover and carefully remove and remeasure the gaging material.

5. The clamp is equal to the split-line shim thickness less the measured gaging material thickness. For example,

$$\text{Splitline shim thickness} = 0.0050 \text{ in}$$

$$\text{Original gage thickness} = 0.0100 \text{ in}$$

$$\text{Measured gage thickness after installation} = 0.0045 \text{ in}$$

$$\text{Clamp} = 0.0050 - 0.0045 = 0.0005 \text{ in}$$

6.5   Bearing Shell Alignment

Bearing shells will be either non-self-aligning (cylindrically seated) or self-aligning (spherically seated), depending upon the length of the bearing and intended service.

1. *Non-self-aligning bearings:*   Generally these are short bearings. These bearings have seats which are cylindrical and roll into and match a cylindrical fit in the bearing housing. To prevent edge loading, the tolerable limit of misalignment between a bearing liner and journal is about 0.00024 in per inch of axial length.

2. *Self-aligning bearings:*   These bearings are characterized by spherical surfaces at the outside of the bearing shell and the inside of the bearing housing. This special feature may be machined into the shell itself or into removable pads that additionally permit shimming of the bearing to adjust rotor height or position. The advantage of a spherically seated bearing is that it can be shifted at the time of installation into perfect alignment with the journal. The bearing does not shift to accommodate transient misalignment in operation. Loose-fitted bearings can shift and cause vibration. Therefore, a clamp of metal to metal or slight interference (up to 0.002 in) fit is recommended.

6.6   Types of Bearings

Rotors supported by oil-film bearings are subject to instabilities in the oil film itself. The resulting vibration can be catastrophic. A basic difference exists between vibration due to instability and vibration due to other faults such as imbalance. In the case of shaft imbalance, vibration of the shaft is a response to the imbalance force occurring at running speed and is proportional to the size of the force. Instability, however, is a self-excited vibration that draws energy into vibratory motion and is relatively independent of rotation frequency. This self-exciting action in the bearing oil film is called *oil whirl* or sometimes *oil whip*. It usually occurs at 43 to 48 percent of the frequency of the rotating shaft.

Stability is normally maintained through damping in the rotor-bearing system. There are numerous bearing designs. Listed in order of increasing stability (i.e., the plain journal is the least stable and the tilting pad journal is the most stable) is a brief description of some of the designs. Increased bearing stability is obtained at higher manufacturing costs. The antiwhirl bearings all impose a parasitic load on the journal which causes higher-power losses to the bearings and in turn requires higher oil flow to cool the bearings.

6.6.1 Plain journal

The bearing is bored with equal amounts of clearance (on the order of 1½ to 2 mils per inch of journal diameter) between the journal and bearing. The bearing normally has two axial grooves at the parting line to distribute the oil (Fig. 6.9).

6.6.2 Pressure dam journal

This bearing is a plain journal bearing with a pressure pocket or dam cut in the unloaded half. This pocket is approximately ¹⁄₃₂ in deep with a width 50 percent of the bearing length. This groove or channel covers an arc up to 135° in the direction of shaft rotation and terminates abruptly in a sharp-edge dam. The dam traps oil above the journal and develops a hydrodynamic pressure which is additive to the bearing load carried by the lower shell. The approach is to load the bearing to 150 to 200 lb/in² of projected area. Pressure dam bearings are for one direction of rotation only (Fig. 6.10).

6.6.3 Multilobe journal

The most common multilobe bearing design is the three-lobe symmetric bearing consisting of three 120° arcs whose centers are displaced from the bearing center (Fig. 6.11).

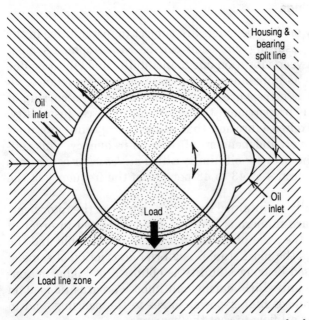

Figure 6.9 Plain journal bearing—two axial grooves—load zones.

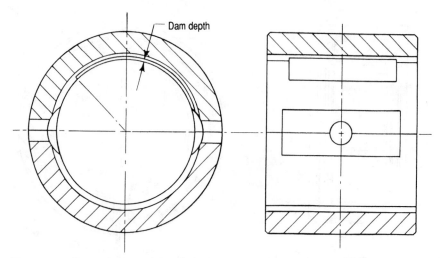

Figure 6.10 Pressure dam journal bearing.

6.6.4 Tilting pad journal

The tilting pad bearing consists of three or more pads located around the circumference (Fig. 6.12). Each pad is able to tilt in order to assume the most effective working position. This design is considered to offer the most positive solution for elimination of shaft vibration attributable to oil whip and oil-film whirl. Pivot action may be accomplished for each shoe by a pin or dowel parallel to the journal axis or by a button.

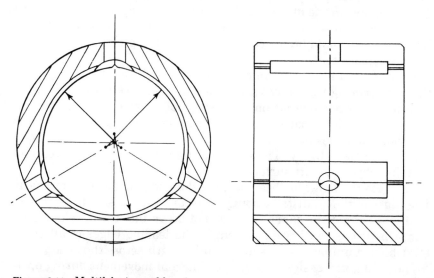

Figure 6.11 Multilobe journal bearing.

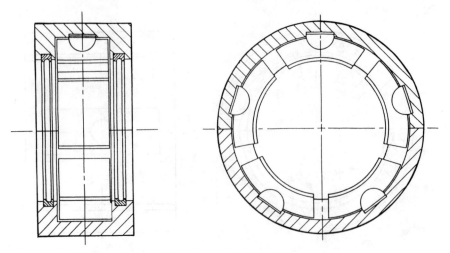

Figure 6.12   Tilting pad journal bearing—spherical pivot.

The first three designs are closely related. The tilting pad design is completely different in concept and has seen general use in large pumps only in the last decade.

6.7 Inspection of Journal Bearings

The three major bearing life factors are the adequate supply of oil to the bearing, temperature control, and cleanliness of the lubrication system. Water in the lubricant oil is extremely dangerous. Even small amounts can cause pitting and corrosion. Acid is equally dangerous. Lubricant oil should be tested frequently for the presence of water as well as acid content.

Moisture and relatively high temperatures of the lubricant cause a tin oxide coating that may vary from a molted dark gray to a highly polished black depending on the degree of overheating and the oil decomposition. The coating builds up to reduce bearing clearance. It is very hard and does not have the desirable properties of babbitt. Scoring of journals is likely. The hard coating can be removed by careful scraping of the babbitt surface.

On pressure-dam-type bearings, the dam edge must be kept sharp. The edges of the oil distribution grooves by contrast should be "feathered" out so that the sharp edges do not break into the developing oil film. On tilting pad journal bearings, the leading edges of all shoes must have a uniform radius for the full length across the shoe.

The tilting pad design experiences internal movement during operation that results in wear at locations other than the area of the oil film proper. As a result, five locations within the bearing should be checked for wear during inspection:

1. Babbitted surface of the shoe.
2. Pivoting surfaces of the shoe and the seat in the retaining ring for most designs.
3. For those designs utilizing the ball pivot, check for wear of the socket and ball.
4. Bore of the seal ring or endplates.
5. The shoe thickness at the pivot point or across the ball and socket— all shoes should be within 0.0005 in of the same thickness.

6.8 Bearing Clearances

The maintenance of cylindrical bearings and tilting pad designs requires different techniques. For example, cylindrical bearing bore measurements should be taken at five positions on both ends of the liner. The measured shaft journal diameter must be subtracted from the measured bearing bore dimensions to determine the clearance in the bearing assembly.

In a tilting pad design, it is necessary to manufacture a stepped-diameter stub mandrel whose smaller diameter is the journal diameter plus the minimum desired clearance (about 0.0015 in per inch of shaft diameter) and whose larger diameter is the journal diameter plus the maximum desired clearance (about 0.002 in per inch of shaft diameter). Mandrel length must be about 1½ times the length of the assembled bearing on each step.

6.9 "Wire Wooling" in Journals

This "strange but true" mechanism is the cause of many machinery failures each year. Journal bearings are often found to have severely scored the shaft without loss of lubrication. The failure mechanism is labeled under the two terms *black scab* and *wire wooling*. The exact metallurgical process is not clear, but in general an iron or chrome carbide particle is formed in the high-chrome steel shaft journal area or thrust collar. This particle "migrates" to the surface, becomes embedded in the soft bearing surface (as it should under bearing design philosophy), and forms a hard deposit of carbide material, or "black scab." The mating steel journal surface is literally machined away by the carbide particle with the formation of fine steel cuttings called *wire wool*, the British term for steel wool. This process is self-propagated when started, and susceptibility to "scab" formation appears to depend upon the nature of the lubricating oil compounding and the metallurgical composition of the steel shaft. Moisture in the oil has also been cited as a cause. The moisture reacts with compounds in the oil to form chemicals that cause increased molecular movement in the shaft steel.

Steels containing chromium or magnesium appear to be susceptible to black scab formation. Shafts of 17-4 PH steels are highly susceptible. Experience indicates that chrome contents as low as 1 percent can cause the formation of the carbides. Chrome content in a shaft above about 7 percent seems to be a sure thing for formation of black scab.

The problem can appear suddenly after several years of trouble-free operation. Regardless of the initiating mechanism for the carbide migration, there are three ways of avoiding the wire wooling type of failure.

1. Do not use high-chrome-alloy shaft chrome content material unless absolutely necessary. It is not known if 1, 3, or 7 percent chrome content is too high.

2. If the high-chrome-alloy shaft is necessary, thermally spray the journal area with chrome oxide or tungsten carbide, which acts as a barrier to the migration of chrome carbide out of the shaft and into the babbitt material.

3. For thrust bearings, use a collar of unalloyed, low-carbon steel, heat-treated to a Rockwell C hardness of 34 to 35.

6.10 Babbitted Thrust Bearings

The amount of thrust developed in a multistage pump is sometimes more than the ball bearings are capable of handling. Babbitted face thrust bearings, shown in Fig. 6.13, must be used to absorb this heavy axial force. The most common thrust bearing is the tilting pad design. The vital elements of a tilting pad thrust bearing are the rotating collar and stationary shoes. The shoes are pivoted to allow them to assume slightly different angles to accommodate an oil film. Oil is introduced at the inner edges of the shoes so that the centrifugal action of the rotating collar causes the oil to flow radially outward. At the same time, the oil adhering to the surface of the collar moves circumferentially and builds up a pressure in a wedge-shaped oil film between each shoe and the collar (Fig. 6.14). There are two basic thrust bearing types.

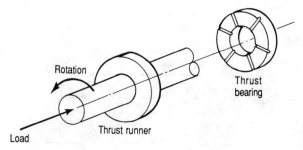

Figure 6.13 Basics of babbitt faced thrust bearing.

6.10.1 Tilting pad without leveling plates

The tilting pad bearing is composed of a series of individual thrust pads which are free to pivot. The pivot line is radial so that each pad can be inclined in a circumferential direction. This bearing is commonly available in 3-, 4-, and 6-pad designs. Using the greater number of shoes reduces the housing size. Some vendors call this bearing a *Michell* after the English inventor. Other companies call it a *Kingsbury type.*

6.10.2 Tilting pad with leveling plates

This bearing is similar to the tilting pad bearing described above, except for an equalizing arrangement which supports the pads and

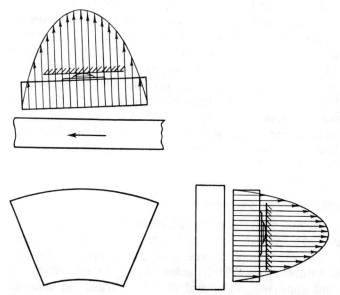

Figure 6.14 Tilting pad thrust bearing oil film pressure profile.

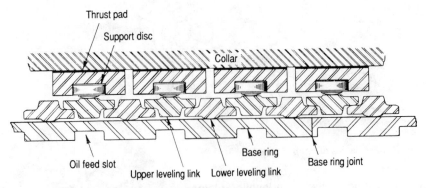

Figure 6.15 Self-aligning tilting pad thrust bearing.

allows load equalization to a great extent (Fig. 6.15). This bearing is generally called by the generic term *Kingsbury*. The allowable loading for the design is about 250 to 450 lb/in^2. The actual values will depend on specific design features, materials, rotational speed, etc.

6.10.3 Thrust bearing maintenance

The wedge-shaped oil film plus minute misalignment of the thrust collar and bearing pads causes movement and wear of the various bearing parts. Heavy thrust loading due to wear on the balancing device accelerates the bearing wear problem. This movement causes wear at seven points in the bearing, as shown in Fig. 6.16.

All these points must be checked for wear. The leveling plates are normally surfaced-hardened. Because the base ring is the softer component, it is likely to show the most wear. Also a flat surface is more easily evaluated. Experience indicates that wear of about 0.006 in on the base ring will cause "lockup" of the leveling plates. Correct, by replacement of parts or a carefully supervised reworking of the entire bearing, any wear in excess of 0.0015 to 0.002 in in the base ring. All shoe or pad thicknesses at the pivot point should be within 0.0005 in of the same thickness. A thorough inspection for wear may prevent a machine failure.

6.10.4 Thrust bearing oil temperatures

For operation at normal or designed thrust loads, the thrust bearing oil outlet should run at a normal operating temperature between 120 and 160°F. The critical metal temperatures of the pad are nearly independent of the oil temperatures. The babbitt faces should not exceed 220°F. A different approach to the control of the thrust pad temperature has been to replace the steel backing (to which the babbitt is

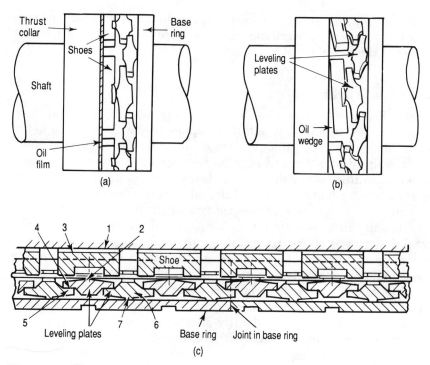

Figure 6.16 Wear points of self-aligning tilting pad thrust: (*a*) stationary view; (*b*) rotating view; (*c*) extended view.

bonded) with a copper-chromium material. This alloy considerably improves the dissipation of the heat generated in the oil film and can reduce bearing surface temperatures by more than 60°F.

6.10.5 How much lubricant oil to a thrust bearing?

Most of the oil flow to a bearing goes to remove heat generated within the bearing by power losses. The causes of power loss in bearings are extremely complex but may be broken down into two parts. The first is the loss due to viscous shear in the oil film between the bearing and collar or journal surfaces. The second is due to turbulence between the exposed moving surfaces of the collar and shaft and the oil.

The shear losses can be determined with reasonable accuracy from normal bearing theory. The losses due to churning of the oil, while negligible at slow speeds, increase rapidly at high speeds. It becomes hard to determine and set the needed oil flows. The rate at which heat is generated in a thrust bearing oil film is roughly proportional to the cube of the bearing diameter. The rate at which heat can be removed from the bearing housing by means of conduction, radiation, and convection is roughly proportional to the surface area, and so to the first power of the diameter of the bearing.

6.10.6 Oil to thrust bearings

Thrust bearings generate considerably more heat than journal bearings. The flow around the pads picks up heat from the collar, and the pads then go straight through to the drain. Only 5 to 10 percent of the oil is dragged into the working film between the collar and the pads. The temperature of the thin oil-film layer between the pads and collar is significantly hotter than the inlet oil temperature. This hot oil recombines with the between-the-pad flow at the periphery of the bearing, usually in a collection chamber or an oil control ring area. The total flow passes onto a central drain where its temperature is measured. Unfortunately, this temperature has little value as a protective measurement, except for detecting serious plumbing mistakes in the bearing supply system. The outlet temperature is not a satisfactory indicator of the thrust bearing condition because it reflects a bulk oil temperature which is only slightly affected by the film temperature. The drain temperature is insensitive to either the load change or the high local babbitt temperature.

Thrust shoe surface temperatures reflect the oil-film temperature and are necessary to accurately protect a machine. Thrust shoe temperatures increase in the direction of rotation, as do fluid-film pressures. Both begin to build after the leading edge of the shoe, reaches a maximum, and begin to decrease as the thrust collar approaches the trailing edge of the pad. API monitoring requirements designate a surface temperature measurement at 75 percent from the leading edge and 75 percent from the bore diameter.

6.11 Lubrication Methods

Sleeve bearings can be lubricated by either oil rings or a pressurized forced feed. When sleeve bearings are utilized in conjunction with a Kingsbury-type tilting pad thrust bearing, both the sleeve bearings and the thrust bearings are lubricated with a forced-feed system.

6.11.1 Ring-oiled journal bearings

Ring-oiled systems are often used for lubrication of horizontal journal bearings. A ring with a diameter larger than that of the shaft rides on top of the journal and dips into the oil reservoir below. The rotation of the journal causes the ring to rotate and carry oil from the reservoir up to the bearing. The top halves of ring-oiled journal bearings are slotted to permit the rings to rest on the journal and deliver oil to the top half. This method of lubrication is adapted only to downward load. When the journal loading is shifted too far from this downward position, the bearing is liable to run warm, because the load commences to fall on the top half of the bearing, which is not designed for this purpose. The same is true when the load falls near the split halves of the bearing. The junction will disrupt the oil film unless the bearings are modified. The general field of application for nonpressurized journal bearings is limited by the speed at which the friction losses are so great that oil cooling must be supplied. Experience indicates that this maximum speed occurs at journal surface velocities of 3500 to 4000 ft/min (3300 r/min for a 4-in shaft) when water cooling coils, water jackets, and air-cooled fins are used and at 2000 to 2500 ft/min otherwise (2600 r/min for a 4-in shaft).

It has been found that

1. The amount of oil delivered to the shaft is roughly proportional to the area of the ring.

2. At high speed the oil is thrown from the ring by centrifugal force on the upward journey, thus necessitating the use of special grooves to collect the oil and deliver it to the shaft.

3. Heavy rings deliver more oil than light ones.

4. Ring diameters of 1.75 to 2 times the diameter of the shaft will provide adequate lubrication for about 4 in on either side of the ring.

5. The oil level in the reservoir must be maintained so that at least 1/4 to 3/8 in of the ring bore is immersed.

6. Oil rings must be perfectly round to function.

6.11.2 Forced lubrication of bearings

In forced-feed systems, the main oil pump can be coupled directly to the pump shaft or separately driven by an electric motor. In addition to the main pump, an auxiliary oil pump can be provided to ensure an uninterrupted supply of oil to the bearings in case of a main pump failure. The driver can be either steam turbine or electric motor. The balance of the system consists of an oil reservoir, cooler, filter, and appropriate pressure switches, gages, and controls. In such a system, one oil pump is normally arranged to provide oil to each bearing before the pump is started. Upon start-up, the auxiliary oil pump is placed in automatic standby service.

Shaft Sealing

7.1 Shaft Sealing

To prevent leakage of the pumped fluid from the pump, the shaft must be sealed where it extends through the pump casing. There are two common ways to seal a shaft to prevent or minimize leakage: packing and mechanical seals.

7.2 Packing

The simplest method of sealing between the shaft and the pump case is to use packing (Fig. 7.1). The stuffing box of a pump provides an area in which to seal against leakage out of the pump along the shaft. It consists of five rings of packing, a lantern ring used for the injection of a lubricating and/or flushing liquid, and a gland to hold the packing and maintain the desired compression for a proper seal.

The function of packing is to control leakage, not to eliminate it completely. The packing must be lubricated, and flow of 40 to 60 drops per minute out of the stuffing box must be maintained for proper lubrication. About 70 percent of the wear on the packing takes place on the first two outer packing rings; so if one fails, another does the job, and the pump need not be shut down.

When abrasive or hazardous liquids are pumped, it is necessary to inject a clean lubricating liquid from an external source (gland oil) into the lantern ring. A flow of 0.2 to 0.5 gal/min is desirable, and a valve should be used for accurate control. The flush liquid pressure should be from 10 to 15 lb/in² above the stuffing box pressure; anything above this will only add to packing wear. The lantern ring is normally located in the center of the stuffing box. However, for extremely thick

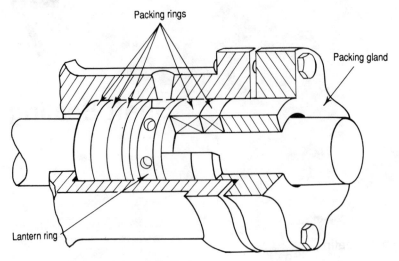

Figure 7.1   Typical packing installation.

slurries it is recommended that the lantern ring be located one or two rings closer to the stuffing box bushing.

7.2.1 Packing size

The best packing results can be obtained only when the right size packing is selected. To determine the correct size, measure the diameter of the shaft (inside the stuffing box area if possible) to give the OD of the ring. All standard packing is made up in increments of $\frac{1}{16}$ in from the smallest size available and up. If the calculated size figures under $\frac{1}{16}$ in, it is generally best to use the next greater size. Thus, if the shaft measures $2\frac{1}{2}$ in and the inside of the stuffing box $3\frac{7}{16}$ in, the size of coil packing to use is $\frac{1}{2}$ rather than $\frac{7}{16}$ in.

7.2.2 Packing materials

Several factors govern which packing to use.

- Type of liquid pumped and its acidity (pH) number
- Stuffing box temperature and pressure
- Shaft sleeve surface speed (feet per minute)

In the past two major materials were used for pump packing in the petrochemical industry: asbestos combined with other materials and metallic foils, aluminum or lead. However, the concern about health problems caused by asbestos has lead to extensive efforts to find substitute materials. Among the available substitutes are plastic and graphite fibers. Carbon and graphite yarns are usually manufactured from a base material of pitch, polyacrylonitrile (PAN) fibers, or rayon fibers. There are three basic grades of this type of packing:

- Preoxidized PAN (acrylic) fibers
- Carbon fibers
- Graphite fibers

There are many different braids or constructions for the filaments, some tight, others loose, etc. The construction is a critical factor in the success or failure of a given packing because it controls the amount of compression, lubrication, leakage rate, etc. Because most braiding equipment operates at high speed and applies high loads, it tends to fracture the yarns slightly during the braiding operation. Plastic coatings such as polytetrafluoroethylene (PTFE) give the individual yarns added strength as well as provide lubrication during braiding. Graphite yarn is normally good to 5000°F, but because of the plastic coating it is limited to 500 to 550°F. Special high-temperature lubricants can be substituted for the PTFE that raises the operating limits as a pump packing to about 800°F. At the higher temperatures a graphite powder, molybdenum disulfide, and other materials may also be used as lubricants.

7.2.3 Correct packing installation steps

Maintenance of a stuffing box consists primarily of packing replacement. However, most packing problems are a direct result of improper installation. The following guidelines should be followed to get the best results from the packing:

1. Remove gland and old packing. If the box contains a lantern ring, make certain that all the packing inboard of it is removed.
2. Inspect the shaft or shaft sleeve for score marks or rough spots. A badly worn sleeve or shaft must be replaced; minor wear must be dressed smooth and concentric. Inspect the lantern ring to make sure that holes and channels are open.
3. Clean the bore of the box thoroughly, and be sure that sealing fluid passages are open.
4. Preformed or die-molded rings are preferred. They ensure an exact fit to the shaft or shaft sleeve and stuffing box bore and will provide a uniform packing density since the molded rings are partially compressed. If the packing to be used comes in a continuous

coil, make sure that the ends are cut squarely on a correctly sized mandrel. Start by installing one end of the ring in the box, and bring the other end around the shaft until it is completely inserted.

5. Using a split bushing packing tool, push the packing ring to the bottom of the box. To prevent the formation of a lip on the packing being seated, use a split bussing that has been bored to the OD of the shaft and turned to the ID of the stuffing box bore. Leave the bushing in place, and replace the gland. Pull up on the gland, and seat the ring firmly and squarely.

6. Repeat this method for each ring, staggering the joints 90°. It is especially important to seat the first few bottom rings firmly; otherwise, the rings immediately under the gland which do most of the sealing will be further overloaded.

7. Install the lantern ring, if used. The proper location of the lantern ring is at the point of seal flush entry into the stuffing box *when the packing is compressed*. This passage can be located with a thin piece of wire.

8. Replace the gland and tighten the gland bolts. Make sure that the gland enters the stuffing box squarely. A cocked gland causes uneven compression and can damage the shaft or sleeve when the pump is put into operation.

9. Keep the packing under mechanical pressure for a short time (say about 30 s) so that it can cold-flow and adjust itself.

10. Loosen the gland, and allow the packing to fully expand with no fluid pressure in the pump. Then tighten the gland and bring it up evenly to the packing, but this time only finger-tight. The packing should be loose enough to let you turn the pump shaft by hand.

11. Start the seal flush to the lantern ring or quench gland. Check for proper flow and pressure.

12. Start the pump. The leakage may be excessive, but do not take up on the gland bolts for the 20- to 30-min run-in. If the packing is too tight, either shut down the pump and repeat steps 8, 9, 10, and 11, or keep the pump running and loosen the gland bolts a couple of flats.

13. If leakage is still more than normal after the run-in period, tighten the gland bolts evenly, one flat or a sixth of a turn at a time. This should be done at 5- to 10-min intervals until leakage is reduced to normal. This may take several hours but will pay for itself many times over in maximum packing and sleeve life.

7.3   Mechanical Seals—General

The most common method of sealing centrifugal pumps today uses the mechanical seal (Fig. 7.2). The three basic parts of a mechanical seal are as follows:

1. *Primary seal components:*   These consist of an axially adjustable sealing ring and an axially fixed mating ring against which it slides, forming a seal. Depending on the seal configuration, either the sealing or the mating ring may be the rotating or stationary element.
2. *Secondary seals:*   These consist of O-rings or other suitable gasket materials.
3. *Hardware:*   This consists mainly of the spring(s) and a retainer, normally made of a series 300 stainless steel.

7.4   Seal Design—Types of Mechanical Seals

Mechanical seals can be classified in two basic types: pusher type, in which the secondary sealing members slide along the shaft sleeve as the seal moves and wears, and nonpusher or bellows type, in which the secondary seal is fixed to the shaft.

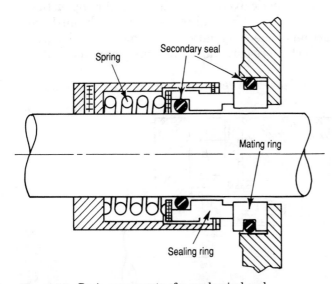

Figure 7.2 Basic components of a mechanical seal.

7.4.1 Pusher-type mechanical seals

The term *pusher,* as applied to mechanical seals, reflects the fact that as any face wear takes place, the sealing element has to be pushed forward along the sleeve to maintain a seal (Fig. 7.3). The pusher type uses a dynamic elastomer as a secondary sealing element that combines both mechanical and hydraulic forces to create and maintain the seal. The more common shapes for these secondary elastomers are shown in Fig. 7.4. The U-cup and the wedge designs are mechanically preloaded to the shaft by means of the springs and do not have the ability to roll or flex as the shaft moves. Consequently, they rub on the shaft and eventually remove the protective oxide surface layer. This ongoing process results in increased shaft wear, scoring, and fretting that eventually result in seal hang-up and the opening of the seal faces.

O-rings, however, due to the differential pressure across them, do not require mechanical spring loading to obtain an effective seal. An O-ring can flex about 5 to 10 percent of its cross-sectional diameter and still roll to make a tight seal. However, like the other secondary seal shapes, O-rings can also fret a shaft sleeve, but the process usually takes much longer.

The sliding action of the secondary seal requires that the surface of the shaft sleeve be machined to a much higher degree of surface finish than for a nonpusher seal. The shaft sleeve should clear the inside of the sliding seal member by 3 to 5 mils radially and should have a finish of 6–16 microinches arithmetic average roughness (R_a) rougher than 16 R_a can abrade the O-ring as the seal moves. This high finish improves the sliding action of the secondary sealing members along the sleeve, but requires much more care during assembly and installation to prevent damage.

A fretted area on the sleeve can result in seal hang-up and leakage.

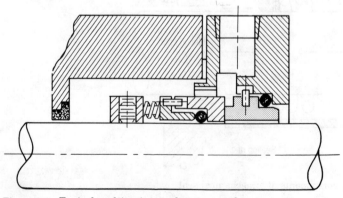

Figure 7.3 Typical multispring pusher-type seal.

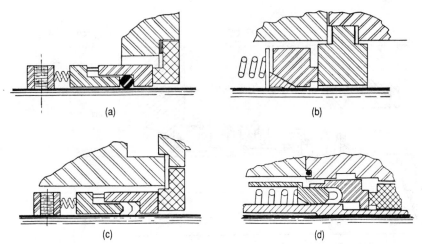

Figure 7.4 Common shapes for secondary elastomers: (*a*) O-ring; (*b*) wedge; (*c*) V-ring; (*d*) U-cup.

The opening of the seal faces due to seal hang-up is usually cyclic; e.g., any axial shaft movement will open the faces and let liquid-entrained abrasives enter. However, the faces usually do not stay open, but close again rather quickly. The delay before closing is called *hysteresis*. In addition to fretting and wear, seal hang-up may aggravated by

- Abrasive particles lodged between the shaft and the sliding components

- Fluids that form crystals in the area where the elastomer rubs on the shaft

- Products that set up or harden around the elastomer

- Solids that become lodged in the springs, preventing the faces from staying in contact when the shaft moves

7.4.2 Nonpusher-type mechanical seals

The secondary seal does not slide along the shaft sleeve in the non-pusher or bellows type of seal; therefore, it is far less susceptible to seal hang-up. Nonpusher-type mechanical seals utilize both elastomeric and metallic bellows depending on the sealing environment (Fig. 7.5). The bellows forms a static seal between itself and the shaft at one end of the seal while the bellows is allowed to float free at the other end. Acting as an accordion, the bellows permits extensive axial movement of the shaft without opening of seal faces. Normally one large spring or several small springs are used to load the rotating face of an elasto-meric bellows.

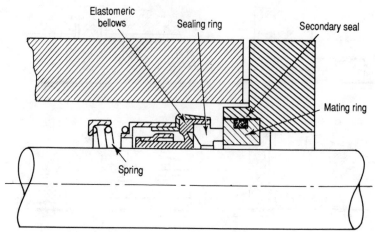

Figure 7.5 Elastomeric bellows nonpusher seal.

Because there is no sliding secondary seal and the elastomeric bellows must seal against the shaft sleeve, the sleeve requires a rougher surface finish (32 to 50 rms). However, to obtain the proper bellows-to-sleeve seal, it is very important to use the right installation lubricant.

The most important advantages of the elastomeric bellows seal are its simplicity and low initial cost. However, the major disadvantage is the limited operating conditions in which it can be applied; the most commonly used bellows materials deteriorate rapidly in the presence of aromatics and are limited to about 200 lb/in^2 and 200°F.

The metal bellows design eliminates the dynamic packing (Fig. 7.6). Instead, a static packing is used to seal the drive collar to the shaft. Because there is no movement between the static packing and the shaft surface, there is no possibility of shaft fretting or seal hang-up. For high-temperature applications, flexible graphite is frequently used as a secondary seal.

The metallic nonpusher mechanical seal uses a series of thin (4 to 7 mils thick), alternating convex and concave, metal disks welded together to form a leaktight bellows. The bellows design eliminates the need for any additional springs as well as any elastomeric gaskets. It is also desirable because it creates a more uniform and constant face loading between the rotating and stationary members, and it more readily conforms to shaft misalignment than spring-type seals.

Hysteresis, while not as prevalent as in the pusher-type seal, can be a problem with the metal bellows seals. However, a more common problem associated with this design is face tracking, or how concentric the rotating flexible element seal face runs against the stationary seal

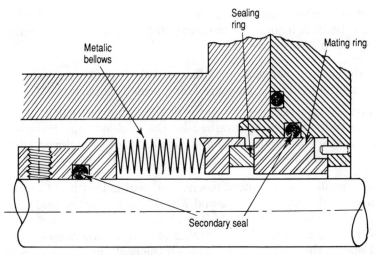

Figure 7.6 Metal bellows nonpusher seal.

face. Since the rotating face is free to float, it tends to run off-center and vibrate. This is a result of imbalance introduced in the manufacturing process and the excitation of the natural frequency of the bellows, which acts and responds dynamically as any other spring. Some eccentric face tracking may actually aid in lubricating and maintaining the seal; however, if it becomes excessive, the faces may open and cause leakage. Tabs on the inside of the seal ring are designed to limit this movement.

The advantage of metal bellows, nonpusher-type mechanical seals is the wide range of applications for which they are suited. They can handle a range of pumping temperatures from a -320 to $+800°F$ and pressures to greater than 300 lb/in^2. The main disadvantage of metal bellows seals is the initial cost.

7.4.3 Balanced or unbalanced seals

In order for a mechanical seal to work, a thin film of fluid must exist across the faces, providing seal face lubrication. However, there must also be some closing force or preloading (sometimes called unit loading) of the seal face against the mating seat seal face. The amount of unit loading determines both the susceptibility of the seal faces to open and the expected life of the seal. If the value is too low, the faces can open easily, allowing leakage, whereas if the value is too high, no film remains between the seal faces and excessive wear and heat will ultimately result in seal failure.

All mechanical seals can be classified as being a hydraulically balanced or hydraulically unbalanced design. The difference is in what happens to the unit loading as the hydraulic pressure against which the seal is holding increases.

1. The unit loading between seal faces of the unbalanced seal increases as the hydraulic pressure being sealed increases.
2. The unit loading between the faces of the balanced seal is established initially and is maintained by the seal springs. However, there is very little increase in that preload as the hydraulic pressure sealed is increased.

The advantage of an unbalanced mechanical seal is that it is the simplest of all seal designs (Figs. 7.2 and 7.3). It allows for the seal to be mounted directly on a straight shaft or shaft sleeve.

However, there are some disadvantages to using an unbalanced seal. As noted above, the unit loading, between the seal faces, increases as the internal pressure being sealed increases. In general, that means that the higher the hydraulic sealing pressure, the higher the unit loading and the shorter the seal life. This effectively limits its application to relatively low pressures. In addition, frequently there are pressure surges associated with starting and operating a pump that must be accommodated by the seal. An unbalanced seal has difficulty handling these surges without the seal faces opening.

For higher-pressure applications, the balanced seal is normally used (Fig. 7.7). A balanced seal can normally be identified simply by looking

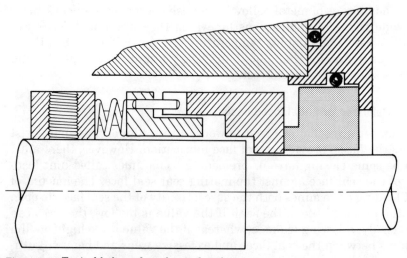

Figure 7.7 Typical balanced mechanical seal.

at it. The most prominent feature is a step on the shaft sleeve underneath the rotating element. In addition, the rotating element has a step on the bore and a step on the outside diameter near the contact faces. Initially at start-up and at low pressures, the majority of the face loading between the rotating and stationary components, of the balanced seal, is provided by the seal springs. However, because of the step changes, counteracting hydraulic forces are set up such that there is very little increase in the seal face pressure as the pressure of the liquid being sealed is increased. The degree of hydraulic balance is commonly called the *seal balance ratio* and is expressed as a percentage of the hydraulic pressure in the seal chamber which exerts a closing force on the seal. The term *balanced* is not really technically correct since typically only about 70 percent of the hydraulic pressure is countered. An additional advantage of the balanced seal is that it will also start and operate under vacuum conditions. The balance ratio is calculated as follows:

- For OD pressurized seals the balance ratio is

$$\frac{(\pi/4)(OD^2 - BD^2)}{(\pi/4)(OD^2 - ID^2)} \times 100$$

- For ID pressurized seals the balance ratio is

$$\frac{(\pi/4)(BD^2 - ID^2)}{(\pi/4)(OD^2 - ID^2)} \times 100$$

where OD = seal face outside diameter
      BD = balance diameter
      ID = seal face inside diameter

For the same size seal, provided the same materials are being used, balanced seals and unbalanced seals have about the same initial cost. The only difference is that balanced seals require a stepped sleeve. The different style sleeve is normally not an additional-cost item since most pumps have sleeves anyway. However, in general, a seal one size larger (approximately $\frac{1}{8}$ in) may be required for a given shaft size if a balanced seal is used.

7.4.4 Seal configurations

The most common seal configuration is the single seal. However, as environmental air quality regulations become more stringent, other arrangements utilizing multiple seals are being used more frequently.

Single seals   A *single mechanical seal* (Figs. 7.5 to 7.7) is a mechanical device (including all associated hardware) having one pair of mating and seat rings, designed to seal against seal chamber pressure which is normally greater than suction pressure. The seal may be lubricated by either the pumped fluid or an external source.

Multiple seals   Normally *multiple seals* are applied when hazardous or toxic liquids are pumped, where a seal failure cannot be tolerated, or when required for fugitive emissions control. The most common multiple-seal arrangements include the tandem and the double seal and a variation of the double-tandem configurations commonly called the *dual seal*.

Tandem seals   A *tandem seal* (Fig. 7.8) consists of two mechanical seals mounted in any orientation (most commonly mounted in series). It utilizes a buffer fluid supplied to the outer seal from an external source at a pressure *lower* than that found in the inner seal chamber. The inner seal works just as a single seal does, normally providing the primary seal to contain the pumpage. The outer seal normally seals only the buffer fluid which also lubricates and cools it. However, in the event the inner seal fails, the outer seal automatically takes over and continues to prevent the pump fluid from reaching the atmosphere.

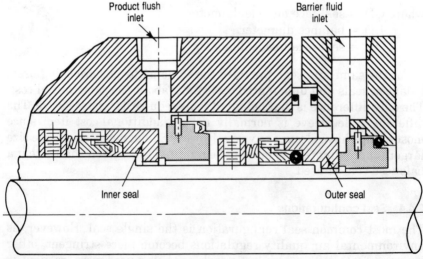

Product flush inlet   Barrier fluid inlet

Inner seal   Outer seal

Figure 7.8   Typical tandem seal arrangement. Barrier fluid pressure < inner seal flush pressure.

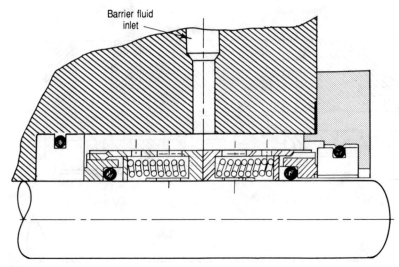

Figure 7.9 Typical double seal. Barrier fluid pressure > seal chamber pressure.

Double seals A *double seal* (Fig. 7.9) is an arrangement whereby two seals are mounted either face to face or back to back. A barrier fluid between the two seals is supplied from an external source at a pressure *higher than* the inner seal chamber pressure. The barrier fluid must be carefully selected since any leakage across the inner seal will mix with the pumped product and any leakage across the outer seal will reach the atmosphere. This arrangement is useful when the pumped liquid contains solids, has poor lubricity, or is otherwise difficult to seal. The biggest disadvantage is that in order to work, the barrier pressure must always be greater than the inner seal chamber pressure; a loss of barrier fluid pressure may result in a pumped-fluid leak to the atmosphere.

Dual seals A *dual seal* (Fig. 7.10) is arranged similarly to the tandem seal in a series configuration and utilizes an externally supplied barrier fluid. However, in this design, the barrier fluid is supplied at a pressure *greater than* the inner seal chamber pressure. In the event of an inner seal failure or a loss of barrier fluid pressure, the outer seal will continue to prevent leakage of pumped fluid to the atmosphere.

Individual seals, shaft sleeves, gland plates, etc., can be purchased for installation in centrifugal pumps. However, the simplest *and* most reliable seal assembly is one that utilizes a *cartridge seal*. A cartridge seal is a complete, self-contained unit, including the seal (single or multiple), gland plate, sleeve, mating ring, etc., which is preassembled, preset, and tested before installation.

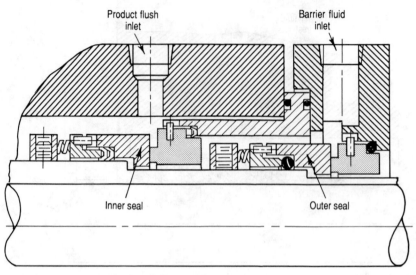

Figure 7.10 Typical dual seal. Barrier fluid pressure > inner seal flush pressure.

7.5 Mechanical Seal Components

7.5.1 Primary sealing members

Sealing faces of mechanical seals are usually only about ¹⁄₁₆ to ⅛ in wide, making the proper material selection and manufacturing techniques used for these components very critical. The seal and mating ring combination normally use a hard material running against a soft material. For process pumps, the most commonly used hard materials are tungsten and silicon carbide while carbon is the most commonly used soft face. These seal components are lapped in order to maintain a thin film of lubricant between the faces without excessive wear. However, only one of these seal faces, the hard face, needs to be truly flat (generally lapped to two to three light bands or about 1/23 millionth of an inch). Because carbon has a low modulus of elasticity it tends to take the shape of the hard face and therefore does not need to be as flat. Even if the carbon has gone out of flat due to temperature or time lapse effects, it will become flat again, in most cases within a short running time (5 to 6 min), when mated against a lapped surface.

7.5.1.1 Carbon elements. At least four main types of carbon are commonly used in mechanical seals. Carbon has been used successfully in seal applications at temperatures as low as −90°C. However the various types do have some upper temperature limits:

Metal-impregnated carbons	
Babbitt-filled	250°F
Copper-filled	480–570°F
Resin impregnated carbons	
Phenolic and graphite impregnation	480°F
Furfuryl resin impregnation	300°F
Electrographic carbons	
Glass-filled	840°F
Carbon graphite	
Oxidizing	525°F
Nonoxidizing	800°F

Where corrosion of the carbon can take place, these upper limits may be severely reduced. Also carbon does not stand up well in caustic service. Except by the manufacturer's part number, the different carbon materials are not readily identifiable visually.

7.5.1.2 Metallic (hard face) elements There is a wide choice of materials for the hard face sealing member. The most important characteristic is the ability to take and maintain a good surface finish under all operating conditions. The following are some of the more commonly used materials.

Ni-resist This nickel-bearing cast iron does not readily take a high finish and does not wear evenly.

Stellite A thin layer of Stellite is placed upon a stainless-steel backup material and polished to a high finish. It tends to heat check (surface cracking) during operation which results in rapid wear of the carbon. For years, stainless steel overlaid with Stellite was the accepted standard in the sealing industry.

Tungsten carbide In the more difficult ranges of pressure and temperature, tungsten carbide has physical properties which improve the performance of seals. Tungsten carbide is harder than Stellite. Its corrosion resistance is quite good in most liquids, and its thermal conductivity is roughly 2½ times greater than that of Stellite overlay. Normally tungsten carbide is an excellent choice as a primary seal member since it dissipates heat rapidly and is highly resistant to the effects of abrupt temperature changes. It can easily be lapped and maintains its high finish as it wears. It is generally made with two different binders, cobalt and nickel. Cobalt binders are prone to corrosion when seawater is present. Cobalt also has an affinity toward copper deposits. Nickel binders are more corrosion-resistant in these services, but the wearing characteristics of the alloy are reduced. Tungsten carbide is extremely hard but is also brittle. A sharp blow can cause it to shatter. It should be handled with care.

Other carbide material Silicon carbide, chrome carbides, and other materials like them are coming into wider use especially in higher-temperature services. The same basic comments that apply to tungsten carbide apply to these materials.

Ceramic Generally a pink or white color is used in special applications because of chemical resistance. Ceramic seals are highly susceptible to sudden, large temperature changes that can cause them to shatter.

7.5.2 Secondary sealing members

There are five or six potential leakage points in the typical mechanical seal, of which at least four are associated with the secondary seals (Fig. 7.11):

1. Shaft sleeve to shaft

2. Rotating seal face to shaft sleeve

3. Stationary element to gland plate

4. Gland plate to pump case

Most secondary seals are elastomers or synthetic rubber in the form of O-rings, wedges, and bellows. An elastomer is manufactured, elastic-like material that can be stretched to approximately twice its normal size, and when released, it will return to its original size and shape (or very close to it). Exposure to high temperatures and chemical attack is the most common cause of elastomer failure. The following paragraphs cover some of the more common elastomers in use today.

Fluoroelastomer This is the most common and best general service O-ring and elastomeric bellows material. Commonly called by the DuPont trade name Viton, it is brown, has a temperature application range from 0 to +400°F, and is good for pressures up to 600 lb/in^2. However, above 400 lb/in^2, a backup ring should be used to prevent extrusion of the O-ring. Viton can be readily cemented and is highly resistant to most hydrocarbons. However, in low-temperature service, such as ammonia, it becomes brittle and should not be used. In hydrocarbon streams having a high aromatics content, it should be limited to 200°F.

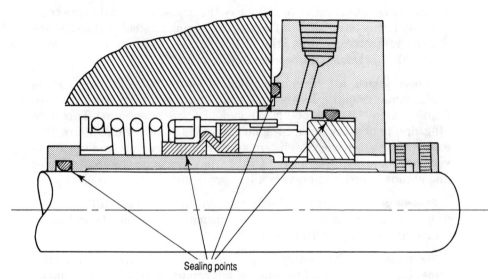

Sealing points

Figure 7.11   Potential secondary seal leak points.

Neoprene   An early-day synthetic rubber, neoprene still remains the best bet in ammonia service after silicone rubber. It is red and has a useful temperature range from 0 to +200°F. It is easily cemented to make O-rings from ribbon stock.

Silicone rubber   Silicone rubber is a very resilient elastomer with a high chemical resistance. It has a rust color and a temperature range of −135 to +450°F. It is not widely available because it requires special molding techniques and is difficult to cement. It is a very good selection for ammonia service.

Nitrile (Buna-N)   This is a widely available synthetic rubber with fair resistance to most chemicals. It is black, and has a temperature range from −40 to +250°F.

Polytetrafluoroethylene (PTFE)   Commonly called Teflon, a DuPont trade name, PTFE is white and has a temperature range of −100 to +400°F. PTFE is also available in a 25 percent glass-filled makeup having a temperature range of −350 to +450°F. Teflon is not a true elastomer in that it has no memory. However, used in conjunction with mechanical seals, it is usually classified as one. Teflon is widely used for wedges, backup rings, and V-gaskets. It does not work well as an O-ring because it does not have sufficient resiliency to function in this manner. Unlike a true elastomer, PTFE allows debris particles in the pumpage to embed themselves in the material surface, frequently resulting in a leakage path. In addition, any sliding action of the rotating secondary seal unit tends to rub away the protective metal oxide coating that forms naturally on the sleeve's stainless-steel surface. Without the protective coating, the sleeve surface loses its resistance to chemical attack and can become highly pitted.

Flexible graphite foil   Frequently called Graphfoil, a Union Carbide trade name, graphite foil is dark gray and has a temperature range from -400 to $+750°F$ ($1600°F$ in nonoxidizing atmospheres). This is an all-graphite product containing no resin binders or organic fillers. It is especially inert at very high temperatures. It is compressed until it is very hard and requires a very good surface finish on the shaft sleeves and glands in order to provide a leakproof seal. It is not an elastomer.

Ethylene propylene (EPR)   A fairly new synthetic rubber, EPR has good properties for use around synthetic lubricants. It is purple and has a temperature range of 10 to $+500°F$.

FFKM elastomer   Known by its DuPont trade name *Kalrez,* this elastomer has good chemical resistance, but compared to other elastomers, it tends to be more expensive and not as readily available. It is black and has a temperature range of 10 to $+500°F$. Caution should be used when existing O-rings are replaced with FFKM in that it tends to swell when exposed to higher temperatures and therefore may require wider O-ring grooves.

Elastomer manufacturers may rate their products for higher temperatures than those given above, and it may be true that these elastomers can tolerate higher temperatures without destroying themselves. However, depending on the temperature and the time duration at that temperature, elastomers will tend to take a compression set and may leak. In addition, some synthetic rubbers (neoprene and Buna-N) deteriorate because of ultraviolet radiation, while others are sensitive to heat and ozone. Storage of all elastomers in a cool dark environment away from electrical apparatus is recommended.

7.5.3 Springs

7.5.3.1 Function of springs in a mechanical seal   The pressure between the two seal faces is partially supplied by a spring. As the seal faces wear, the flexibly mounted seal ring must move along the shaft sleeve toward the mating ring. The total amount of movement in a balanced seal is essentially the distance between the sealing face and the step on the OD of the mating ring (about $1/16$ to $1/8$ in). Spring face loads should vary between 30 lb when the carbon is new and 10 lb at the end of the carbon life. In order for the springs to continue to be able to apply a force on the seal faces, even in the worn condition, the springs or bellows must be set with an initial amount of compression. If specific data are not available, an initial compression of about ⅛ in is a good rule of thumb for most seal installations.

7.5.3.2 How many springs? How many springs should be used in a seal? There are several advantages and disadvantages of the multiple- (Fig. 7.3) and single-spring (Fig. 7.2) seal designs.

Multiple Springs

Advantages:

1. The multiple-spring design has a shorter axial stuffing box space requirement than a single coil spring.
2. Multiple combinations of the same spring can be used with seals of many sizes.
3. Face loading (about 30 lb) can be more readily varied simply by adding or subtracting springs.
4. Multiple springs also resist unwinding from centrifugal forces to a much higher degree than single coil springs, since the forces act differently and the springs are caged in retainers.

Disadvantages:

1. The small cross section of the wire can corrode easily. However, stainless steels and Hastelloy can compensate for much of this disadvantage.
2. The small springs are more likely to clog from debris.

Single Springs

Advantages:

1. Variations in face loading can be attained with a single coil spring, but the rate of deflection is such that it requires a large axial displacement to effect a relatively small force change.
2. A corrosion allowance can be built into the larger-cross-section wire.

Disadvantages:

1. Shaft rotation is a big factor in single-spring seal applications. The spring should be placed in a retainer to resist unwinding. The retainer is large and heavy, introducing a balance and vibration problem. The hysteresis problem can be increased.
2. The unwinding of the spring can be coped with by providing either a right-hand or left-hand coil so that rotation tends to tighten the coil.

Spring material can also be a problem. Stainless-steel springs can fail from chloride stress cracking, a problem caused by a combination of stress, chlorides, and temperature. Remember that, outside of water, there are more chlorides in this world than anything else; salt is a chlorine compound, the most common water treatment is the addition of chlorine, and chlorides are also present in many cleaners in the form of chlorinated hydrocarbons. Hastelloy C spring material is frequently used as a solution to the chloride stress problem.

7.6 Seal Flush Systems

High-temperature fluids containing abrasives are common mechanical seal application problems. The abrasive will generally penetrate between the seal faces and cause rapid wear of the faces. Excessive heat from the pumped fluid or friction will cook the elastomers and distort seal components, causing the seal to leak and fail. A means of providing the seal with a clean, abrasive-free lubricant and of removing the heat generated at the seal faces must always be provided. This can be accomplished by using a seal flush. For flushing we can sometimes use a clean liquid from an outside source. However, the resulting contamination of the pumped product generally makes this type of flush undesirable. Therefore, a sidestream or bypass flush from the liquid being pumped is generally used. If necessary, this fluid flush can be cooled and any abrasive particles removed before it is injected into the seal.

7.6.1 Types of seal flush arrangements

The following paragraphs outline a few of the most commonly used seal flush piping plans as established by API Standard 610, *Centrifugal Pumps for General Refinery Service* (Fig. 7.12).

Plan 11 The flow is from the discharge of the pump directly into the seal cavity. This is used only for relatively cool, clean pumpage conditions. The flow enters the seal chamber at the seal faces, flushes the faces, and flows across the seal back into the pump.

Plan 13 This is the reverse flow from plan 11. Flow from inside the pump is directed outward across the seal faces, out the seal gland, and back to the pump suction. This plan is utilized to prevent vapor buildup at the seal faces and is frequently used on vertical canned-type pumps.

Plan 21 Similar to plan 11, this plan takes clean fluid from the discharge of the pump and directs it to the seal faces through a cooler.

Plan 23 Applied on hot services, this plan utilizes an internal pumping ring to circulate the flush fluid across the seal faces, through a cooler, and back to the seal cavity.

Plan 31 Fluids contaminated with particles that are heavier than the fluid itself can be cleaned up by using a cyclone separator. This plan takes flow from the discharge through a cyclone separator, where the

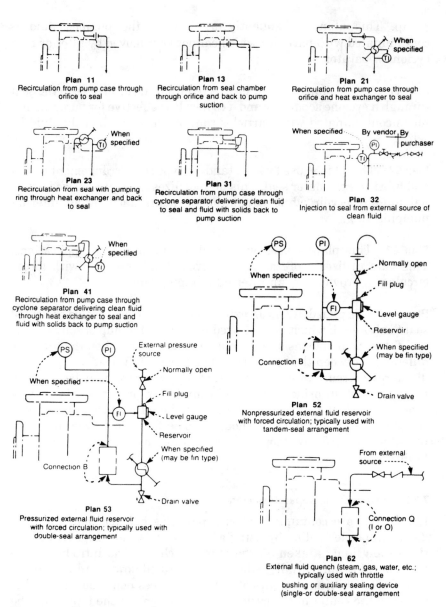

Plan 11
Recirculation from pump case through orifice to seal

Plan 13
Recirculation from seal chamber through orifice and back to pump suction

Plan 21
Recirculation from pump case through orifice and heat exchanger to seal

Plan 23
Recirculation from seal with pumping ring through heat exchanger and back to seal

Plan 31
Recirculation from pump case through cyclone separator delivering clean fluid to seal and fluid with solids back to pump suction

Plan 32
Injection to seal from external source of clean fluid

Plan 41
Recirculation from pump case through cyclone separator delivering clean fluid through heat exchanger to seal and fluid with solids back to pump suction

Plan 52
Nonpressurized external fluid reservoir with forced circulation; typically used with tandem-seal arrangement

Plan 53
Pressurized external fluid reservoir with forced circulation; typically used with double-seal arrangement

Plan 62
External fluid quench (steam, gas, water, etc.; typically used with throttle bushing or auxiliary sealing device (single-or double-seal arrangement

Figure 7.12   Seal flush plans.[1]

dirty fluid is returned to the pump suction and the clean liquid is then circulated to the seal cavity, across the seal faces, and back into the pump.

Plan 32   This plan utilizes a flow of clean fluid from an external source. Fluid is circulated to the seal cavity, across the seal faces, and into the pump. This plan is frequently used when the pumped fluid is contaminated with particles which cannot be removed by filter or by a cyclone separator.

Plan 41   In this plan, the flow is directed from the discharge to the seal through a cyclone separator and a cooler. It is effective for pumping hot fluids contaminated with particles that are heavier than the fluid itself.

Plan 52   This plan is used with a tandem-type seal arrangement to cool and lubricate the outer seal. This plan utilizes a nonpressurized external fluid reservoir while circulation is maintained by an internal pumping ring.

Plan 53   This plan is used with double- and dual-seal applications. This plan utilizes a pressurized external fluid reservoir with forced circulation provided by an internal pumping ring.

Plan 62   This plan delivers an external fluid quench of cooled or heated seal flush liquid, which may be used to prevent buildup of solids at the seal faces. Some means of flow control is necessary when the pressure difference between the various points exceeds 50-ft head of liquid, i.e., discharge pressure to seal chamber, etc. Flow control may be achieved by the use of an orifice plate (⅛-in minimum opening) or a cyclone separator. Whichever method is used, it should be incapable of being completely shut off, and care should also be taken to prevent blockage by sediment since any stoppage of circulation to the seal can cause seal failure.

7.6.2   How cyclone separators work

Fluids that are contaminated with particles heavier than the fluid itself can be cleaned up by using a cyclone separator (Fig. 7.13). Particle-laden fluid is taken off the pump discharge and introduced into the separator tangentially, where it is spiraled downward by centrifugal force. Obeying the laws of nature, the less dense clean liquid is forced to the center of the cyclone and is drawn out the top, while the heavier particles are forced to the outside and drawn off the bottom underflow and returned to the pump suction. For successful cyclone operation, the following recommendation should be followed:

1. The solids content of the pumped fluid should not exceed 10 percent by weight.
2. The pressure-difference (PD) ratio between the inlet and outlet should be from 0.8 to 1.2.

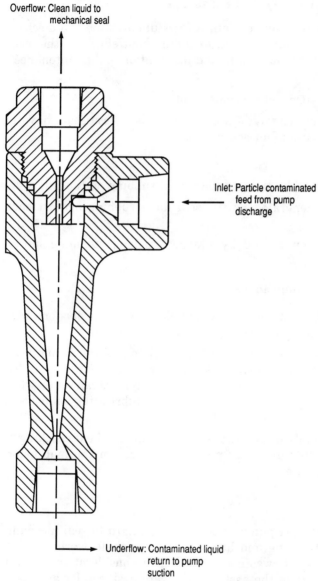

Overflow: Clean liquid to
mechanical seal

Inlet: Particle contaminated
feed from pump
discharge

Underflow: Contaminated liquid
return to pump
suction

Figure 7.13 Typical cyclone separator.

$$\text{PD ratio} = \frac{P_i - P_c}{P_i - P_d} = 0.8 \text{ to } 1.2$$

where P_i = inlet pressure to cyclone
P_c = pressure at "clean" outlet
P_d = pressure at "dirty" outlet

7.7 Pump Design—Why Seal Faces Open

In addition to the basic seal design, successful operation of a mechanical seal depends on many things, including heat effects in the operating environment, pump design, and installation and maintenance.

7.7.1 Heat effects—Operating environment

Heat is the greatest enemy of secondary seals. These seals have to contend with heat from four sources:

- The sealed product
- The heat generated by the secondary seal rubbing on the shaft
- Frictional heat generated by the seal faces rubbing against each other
- Improper operation caused by plugged coolers, steaming out of pumps, etc.

Heat does three things to a seal:

1. It can cook a dynamic secondary seal elastomer. The elastomer becomes hard, cracks, and can no longer seal effectively.
2. Heat can cause the pumped product to change state. Some fluids crystallize between the seal faces and eventually destroy them, hydrocarbons will form coke or gum up, and polymers can set up when exposed to high heat. Other fluids lose their lubricating qualities as they get hot and may eventually flash.
3. Heat can thermally distort the seal parts, causing heat checking or cracking of Stellite and some plated hard faces, or thermal distortion of the seal faces.

7.7.2 Pump design

Many of the older-style pumps were, and are, hard to seal mechanically because of the long and heavy cantilever shaft overhang. The stiffness of pump shafts was not given serious consideration by most pump manufacturers in the past. Pumps were designed for and operated with stuffing box packing rather than mechanical seals. The packing acted almost as a third bearing supporting the shaft, and once the packing is removed, the shaft is free to move around. Reasonable seal life cannot be obtained on pumps with buggy-whip shafts. An important plus for stiffer shaft construction is that less internal wear occurs on wearing rings, bushings, and sleeves. Consequently, operating and maintenance costs are reduced.

Pump cavitation and recirculation problems can cause the shaft to move excessively both axially and radially as well as create hydraulic pressure surges that adversely affect the mechanical seal. Many seal failures are the result of hysteresis or a delay or lag as the seal faces attempt to move. This effect lets the seal faces separate, allowing the entry of abrasives and other solids. When the faces close, the solids embed themselves into the softer carbon face and cause it to act as a lathe cutting tool, which will score and eventually destroy the hard face.

An important consideration in mechanical seal selection is the pump speed. Seal size and pump speed go together. High peripheral speeds create high thermal loads by generating heat at the faces of conventional seals. Shaft size is also a factor since it has a direct influence on the velocity of the seal faces and the amount of heat generated. Another reason for the speed limits involves the vibration of the rotating seal head. Strangely, the seal is the only rotating part of a pump that is not dynamically balanced.

7.7.3 Installation and maintenance:
Installation checklist

Respect the primary (sealing faces) and secondary (O-rings, wedges, bellows, etc.) sealing areas. Do not scratch, nick, handle, or get those areas dirty. Sometimes even a fingerprint can cause a seal to leak. Remember that rubber, Teflon, and gaskets can easily become cut and torn with rough handling.

1. The following parts should be carefully checked before seal installation.
    a. *Reference mark:* Record the distance from the face of the rotating element face to the face of the stuffing box. If the rotating element face is worn excessively, measure the distance from the sealing washer shoulder to the face of the stuffing box.
    b. *Shaft sleeve:* Ensure that the shaft sleeve is free of burrs and sharp edges. It must be finished to the appropriate smoothness for nonpusher or pusher seals. The sleeve must run true when mounted on the shaft. Check the shaft sleeve dimensions carefully. Make sure there is a $\frac{1}{16}$-in by $\frac{3}{32}$-in chamfer at the step on the sleeve over which the seal is to be mounted. Check the finish carefully in the area of the secondary seal and the sleeve-to-gasket O-ring. Many vertical pumps have cartridge sleeve assemblies with an O-ring seal area on the shaft approximately 5 or 6 in below the stuffing box face. This shaft seal area must be clean and have a good seating area. When a sleeve having an O-ring seal in its bore is to be assembled on a threaded shaft,

wrap a piece of shim stock or cellophane tape around the threads, thereby permitting the O-ring to pass over the threads without damage.

c. *Gland:* Check the gland or stationary seat bore of the pump head. A $\frac{1}{32}$-in radius should be at the entrance to the seat bore. It is necessary to maintain this radius to prevent damage to the seat O-ring. The secondary seal area should not be corroded and should not exceed 1-mil total indicator runout with the gland or pump head. Determine which of the pipe threaded holes on the gland circumference are to be connected to the flushing line, quench line, and drain line. The gland should be checked for concentricity. The blowout bushing should be tight in the gland and should have approximately $\frac{1}{32}$ in of diametral clearance to the shaft.

d. *Stuffing box:* The bore of the stuffing box should be concentric within 3 mils so that the lip on the gland (where used) will have a good spigot fit. The throat bushing or restriction built into the entrance of the stuffing box should fit fairly tightly and should have a diametral clearance of about $\frac{1}{32}$-in.

e. *Bearing:* Bearings should be in good condition. Ball bearings should not show any signs of roughness. The shaft should be mounted in the bearing housing; then, by using a dial indicator mounted on the shaft, the following checks should be made before the seal is assembled:

(1) End play or axial movement should not be greater than 3 to 4 mils.

(2) Shaft deflection or runout should be 2 mils maximum.

(3) The drive ring of the seal should be absolutely square with the shaft.

(4) The stuffing box face runout must be within 2-mil total indicator runout.

2. After the preceding checks have been made, the following should be observed during assembly of the seal members:

a. Seat installation

(1) When an O-ring seat is used, oil the O-ring on the seat and press into the bore. Press firmly (do not hammer) until the seat is all the way to the bottom of the cavity in the gland.

(2) When a clamped stationary seat is used, assemble the seat with the gland.

b. Rotating members

(1) *Nonpusher type:* Oil the ID of a bellows, and assemble the seal on the sleeve so that the tail of the bellows is over the shoulder on the sleeve from $\frac{3}{8}$ to $\frac{1}{2}$ in. When the bellows is assembled, compress the seal spring about 1 in and tie the

assembly together with three wires. Then the tail of the bellows may be seated on the shaft at its proper location.

(2) *Pusher type:*   Oil the ID of the wedge ring, U-cup, etc., and assemble the seal on the sleeve.

3. During final assembly of the seal, these points should be checked:
   a. Where a gland is utilized, it should be made up with a feeler gage in order to maintain its squareness with the face of the stuffing box within 2 mils.
   b. Gland bolts should be tightened very uniformly. Overtightening will distort the gland and the stationary washer. This connection can be spotted in a seal that has been in service by the fact that the stationary seat seems to have a high and low point in its wear patterns.
   c. Be sure the seal is compressed the correct amount. A rule of thumb is to compress the seal the length of the step machined on the carbon, plus a little more; but that idea will not work with many seals. So look at a drawing of the seal for an exact dimension.

4. During installation of the pump, check these points:
   a. Make sure that flush connections are properly hooked up. Consider how the pump is built to determine which piping setup is correct. There should be no valves in the flushing line. If a cyclone separator is not used, a restriction orifice should be used (1/8-in minimum opening).
   b. Bleed all air and vapors from the pump prior to and during starting of the pump.
   c. Pump alignment is critical to seal operation. Check it before and after the blinds are pulled.
   d. A vertical pump must have its mechanical seal removed and an aligning washer installed in the stuffing box when the electric motor shaft is aligned with the pump shaft.

Always follow the pump and seal manufacturers' specific instructions and drawings. And during all stages of assembly, *think*.

References

1. American Petroleum Institute Standard 610: *Centrifugal Pumps for General Refinery Service*, Seventh Edition, February, 1989. Figures D-2 & D-3.

Mechanical Considerations

8.1 Lateral Critical Speeds

A critical speed of a pump is any rotational speed which corresponds to
a natural frequency of the pump rotor. Such a critical speed is experienced when an imbalance creates a vibration whose frequency is equal
to a natural frequency of the pump rotor. The term *stiff shaft* is sometimes used to describe a pump whose operating speed is below the first
critical speed. The term *lateral critical speed* describes the critical deflections and mode shapes which occur along the length of a *flexible
shaft*. The dynamic forces exerted during pump operation are not yet
satisfactorily understood, especially with respect to frequency. Turbulent flow, suction recirculation, interaction of impellers and casing,
cavitation, hydraulic imbalance, and the forces from leakage points
such as wear rings and bushings tend to suppress the natural frequencies when the pump is operating. Only during start-up or after extreme
wear at the leakage points does the rotor exhibit its natural frequencies. Calculation of the hydraulic effects on the natural frequency of a
multistage pump rotor is almost impossible. Some manufacturers use
terms such as *wet critical,* which considers the hydraulic forces, and
dry critical, which neglects the hydraulic forces, to describe their rotors. Almost all single-stage pumps operate below the first natural
"dry" frequency. Most multistage pumps operate between the first and
second dry critical speeds.

In addition to lateral critical speeds, rotors can experience *torsional
critical speeds*. Torsional critical speeds are a concern primarily with
multiple-component drive trains. It is important, however, to consider
the possibility of torsional critical speed problems in any large drive
train comprised of three or more elements and two or more flexible
couplings, where varying degrees of flexibility (or variable stiffness)

could result in susceptibility to torsional excitement. When involved with applications requiring multiple pump/drive trains, a torsional critical speed analysis should be performed to ensure trouble-free operation.

8.2 Dynamic Balancing of Components

Precision balancing of pump impellers and rotors requires the removal of metal from heavy areas or the addition of weight (by welding or counterweights) to light areas. The objective is to make the mass center of the rotor coincide with the geometric center about which rotation occurs. The only way to get a rotor with predictable dynamics is to maintain near-perfect balance of each individual component during all operations of machining and assembly. This balance should never be disturbed by making indiscriminate corrections on the finished rotor assembly.

8.2.1 Imbalance limits

Two major residual imbalance limits currently used are roughly compared in Fig. 8.1. The units used are gram-inches (g·in) per 100 lb of rotor weight, to permit a more direct comparison.

1. *American Petroleum Institute (API) Standards 610 (7th edition), 612 (3d edition), and 617 (5th edition):* This imbalance limit standard has been used by the U.S. Navy (U.S. Navy, MIL-STD-167, Ships) for over fifty years and was adopted by API in the late 1980s.

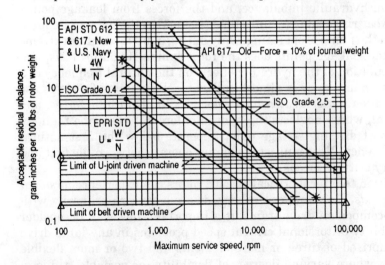

Figure 8.1 Comparisons of imbalance limits. U = unbalance, in gram-inches; W = rotor weight, in pounds; N = rotor operating, in rpm.

It is more restrictive than the older API specifications and much tighter than the International Standards Organization (ISO) standard. The specification states that the residual imbalance in each correction plane shall not exceed

$$\text{Imbalance} = \frac{4W}{N}$$

where N is the maximum actual running speed, in revolutions per minute, greater than 1000 and W is the static journal weight on each end, in pounds.

2. *International Standards Organization DR 1940:* This is a direct plot of residual imbalance per unit of rotor mass versus service speed. Turbine and compressor rotors are assigned a grade of 2.5. The upper limit of this grade is about $15W/N$, while the lower limit is about $6W/N$. Pump impellers are assigned to grade 6.3, which is about $35W/N$. These limits are not satisfactory for many turbomachinery applications. Putting the API and ISO imbalance limits on the same footing would require using about an ISO grade 0.7, as shown in Fig. 8.1.

Several manufacturers object to the tighter API limits as being unnecessary and too expensive. A good balancing machine operator can achieve these balance limits with the expenditure of about 1 more hour of labor or a cost of $50 over the ISO limit. The tighter limits result in a reduction of stresses and forces acting on the machine components, especially the bearings, and make the extra hour of labor the best investment made during the repair of the rotor.

8.2.2 Impeller-balancing procedure

Dynamic balancing of pump impellers is generally done by grinding away metal. The location and manner in which the metal is removed sometimes do not receive the attention they should. The strength of a rotating impeller can be adversely affected by improper grinding. The hub, a highly stressed part, should be avoided. Ground areas should be spread as much as is practical, and sharp corners or deeply ground holes should be avoided. The quality of any dynamic balancing operation depends upon the control of radial runout and the elimination of internal couples along the length of the rotor.

Individual balancing of impeller(s) is vital. The following procedures are a *must:*

1. Prepare half-keys as required for the balancing of individual impeller(s) on a mandrel. These keys must precisely fill the open key-

ways at the impeller bores, similar to the keys used at final assembly.

2. A precision balancing mandrel for impeller(s) must be made. A typical design is shown in Fig. 8.2. The actual geometry should match the minimum pedestal spacing and the roller configuration of the balancing machine. Mandrel design guidelines are as follows:

   a. The mandrel should be made of a low-alloy steel, such as AISI 4140 or 4340, which has been stress-relieved.

   b. The journal surfaces should be hardened and ground, with a finish not poorer than 16 rms.

   c. All diameters must be concentric within 0.001-in total indicator reading (TIR).

   d. The diameter of the section in which the impeller is to be mounted should be established on the basis of heating the impeller hub to approximately 300°F for installation and removal.

   e. Keyways are not incorporated in the mandrel.

   f. The impeller-balancing mandrel should be checked to ensure that it is in dynamic balance. Make corrections on the vertical faces only.

   g. The impeller-balancing mandrel should be free of burrs and gouges.

   h. Mount each individual impeller, together with its half-key, on the balancing mandrel by careful and uniform heating of the impeller hub, using a rosebud-tip torch, to approximately 300°F. A temperature-indicating crayon stick should be used to monitor the heating operation. Cool the impeller by directing a flow of shop air against the hub. When the impeller and mandrel have cooled to room temperature, install the mounted impeller in the

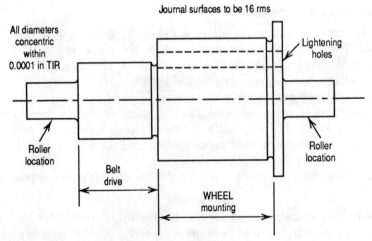

Figure 8.2 Precision balancing mandrel.

balancing machine. Identify the required dynamic corrections with the balancing machine operating at the highest speed for the impeller diameter.

i. Make the required dynamic corrections to the impeller by removing material over an extended area with a relatively fine-grade grinding disk. Blend in the ground area with adjoining contours. Under no circumstances should holes be drilled in the impeller for balancing.

8.3 Process Pump Gaskets

With the banning of asbestos as a gasket material, some serious problems have arisen because no substitute for asbestos has demonstrated total superiority to date. The major deficiency in nonasbestos gaskets is the long-term higher-temperature stress relaxation. This tendency means frequent bolt tightening and the risk of gasket blowout.

8.3.1 Horizontally split pumps

A wide variation exists in the quality and ruggedness of available horizontal casing joints. These variations exist between vendors and between models from the same vendor (in some cases, variations exist between different sizes of the same model or manufacturer). The normal practice for these pumps is to machine the bore with a 0.030-in shim between the halves of the casing. The shim is discarded after boring. A $\frac{1}{32}$-in compressed sheet gasket is used between the flanges. This material has a thickness of 0.030 to 0.038 in as manufactured with a compressed thickness of 0.025 to 0.032 in, a variation of up to 0.013 in. With some asbestos-substitute gasket material, the compressed thickness can be less than the design amount.

The gasket is unconfined. Therefore the proper bolting procedure and sequence must be followed to ensure proper sealing, particularly on multistage pumps and for high-pressure applications. The tightness of a horizontally split casing joint can be lost if the pump is thermally shocked (sudden entrance of a hot fluid into a cold pump), and as a result, the unconfined gasket can be blown out, causing a major leak. The tendency for the stress relaxation of the asbestos-substitute materials increases this possibility.

8.3.2 Vertically split pumps

Some limitations are imposed by API Standard 610 that should be taken into account in selecting pumps.

1. Vertically split pumps are required for pumping temperatures above 400°F (a lower limit should be used if thermal shock is probable).

2. Vertically split pumps are required for flammable or toxic service with specific gravities less than 0.07.

3. Vertically split pumps are required for flammable or toxic services at rated discharge pressures greater than 1000 lb/in^2 gage.

Vertically split pumps use metallic spiral-wound confined gaskets between casing sections whereas the gasket on a horizontally split joint is not confined. Thus, the probability of achieving a proper, continuous seal is greater with a vertically split casing design. In addition, the conversion of the filler in this gasket design to nonasbestos materials has had less impact on the performance of the gasket than the sheet gasket material has.

Many single-stage process pumps incorporate the back pullout feature so that the pumping element may be removed from the case without breaking the piping connections. This construction also permits centerline mounting of the casing to minimize thermal expansion problems, allowing operation at higher pressures and temperatures for hydrocarbons. There is one fact that cannot be overlooked in working on this type of pump. *The head must be made up square.* This is important for two reasons: (1) to obtain correct compression of the gasket for pressure-holding capabilities and (2) to maintain correct internal alignment and positioning of the rotating element with respect to the stationary casing components.

Starting in 1981, the API Standard 610 specification required that the heads be machined such that when the heads are pulled "metal to metal," the gasket will be compressed the correct amount. Pumps made before 1981 may have a separation at the heads when the gasket is made up properly.

8.3.3 Compression of the gasket

Spiral-wound pump head gaskets consist of V-shaped preformed plies of metal, wound up in a spiral with a soft separation of nonmetallic fiber (Fig. 8.3). The V shape gives springlike characteristics as the metal and soft filler plies flow into gasket surface finish irregularities to give sealing action. The inner and outer metal-to-metal plies must be under equal compression. The compressibility of a spiral-wound gasket is controlled for a specific bolt loading of 30,000 or 45,000 lb/in^2 by varying the number of metal-fiber wraps. The standard gasket is good for temperatures up to 750°F and should not be reused. Compression to a predetermined thickness makes a spiral-wound gasket give its best performance. Most pump head gaskets are 0.175 in thick and should be compressed to 0.130 ± 0.005 in.

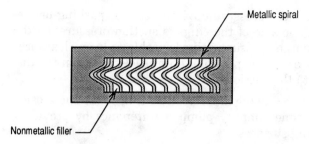

Figure 8.3   Cross section of spiral-wound pump head gasket.

Since the ID of a spiral-wound gasket decreases during compression, there must be clearance of $\frac{1}{16}$ in on the ID to permit equal compression of all plies. The OD can fit in the head recess more snugly, but a tight fit cannot be allowed.

8.4   Impeller Attachment Methods

Nearly all impeller fastening devices for single-stage pumps are threaded in a direction that is counter to pump design rotation, to reduce chances of loosening when the pump is driven in the correct direction. Some impellers are screwed onto or into the end of the shaft. The screwed impeller is particularly vulnerable to backward rotation or backflow of the product. No pump manufacturer has a universal fastener which is used on all pump sizes. The actual fastener method is often not clear in the pump drawing. The fasteners, in general, are run-of-the-mill items that are purchased by the pump manufacturer in bulk. These standard fasteners can cause problems in a process pump. The material can be too hard to be compatible with many of the pumped products (chlorides, etc.). The fastener can then break easily during a run. The same problem is present with lockwashers and cotter keys.

8.5   Wear Rings

8.5.1   General

Wear rings on centrifugal pumps have four primary functions:

1. They provide a running joint between the stationary and rotating parts of the pump.
2. Wear rings provide a means of controlling or minimizing the hydraulic thrust of the pump, as discussed in Chap. 4.

3. They provide a reduced-pressure environment for packing or mechanical seal. On one side of the rings is suction pressure; on the other side is discharge pressure. It is desirable to keep clearances between the rings as low as possible to limit leakage from the discharge side back to the suction side.

4. Wear rings protect the casing and impellers from wear and erosion and at the same time simplify pump maintenance by providing replaceable wearing surfaces.

8.5.2 Mounting of impeller wear rings

Impeller wear rings are shrunk on the impeller and locked in place to prevent movement due to rubbing or a partial seizure. Interference between impeller wear rings and the impeller is 0.001 to 0.0015 in for ring diameters ranging from 2½ to 6 in; 0.002 to 0.0025 in is used for rings 6½ to 12 in in diameter. Two methods of locking the ring are in general use. One method is to drill and tap either radially or axially for screws at the joint between the impeller and the impeller wear ring. After the screw is tightly bottomed, it is sawed off, peened, and dressed down with a fine file. API 610 specifications limit the diameter of the pin to one-third the width of the ring. This method has some problems. Threads must be tight, which causes the wear ring to bulge at the pin with soft materials or may cause hard surfacing to crack. The second method substitutes tack welding for the screws. Thermal stresses and metallurgical changes in the impeller metal must be avoided. API specifications frown on this method, but when properly done, it is a good method. After the rings are locked in place by any method, they should be checked for radial runout.

8.5.3 Mounting of casing wear rings

In most horizontally split pumps, the casing wear rings are mounted with a tongue-and-groove joint, as shown in Fig. 8.4. The wear rings "float" under hydraulic pressure of leakage to a running position and are locked in place by the discharge pressure. The critical dimensions are on the bore and vertical flange face. There are three reasons for this practice:

1. It makes internal alignment of the casing and the rotor easier.

2. The case gasket can be compressed by different amounts without deforming the case wear ring. As discussed in Sec. 8.3.1, the compressed thickness of the case joint gasket can vary by up to 0.013 in. The case cover must be put down "blind," so fits must be "sloppy." In

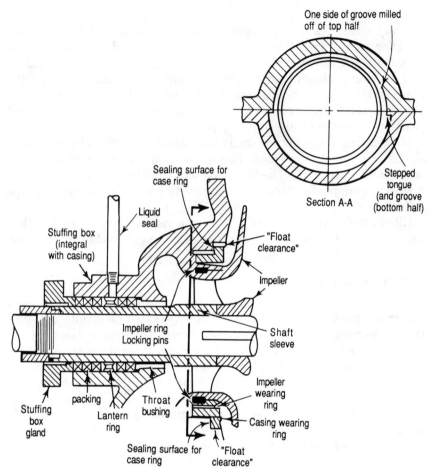

Figure 8.4 Wear rings in a horizontal pump.

Fig. 8.4, the top half of each ring is shaped differently to help prevent any binding of the rings as the cover is lowered into place. Note the antirotation lugs. The "float" clearance should be enough to accommodate about one-half of the variation of compressed gaskets. On smaller pumps it should be 0.004 to 0.006 in on the diameter of the rings. In a large circulating water pump, the clearance may be a little larger. In no instance should the case rings be clamped when the cover is tightened.

3. The seal area between the case and the wear ring is the vertical face of the ring. The case ring is frequently bored out and reused. The impeller ring is replaced.

In vertically split pumps, there are two ways of getting the proper wearing clearance during repairs. One is to install new casing and impeller wear rings. The more economical, second alternative is to remachine the impeller wear ring by mounting the impeller in a lathe and making a cleanup cut on the circumference. After turning, the impeller ring should be measured and a new casing ring machined to dimensions for proper clearance. This is the reverse of normal renewal practices in horizontally split pumps.

Most casing rings for vertically split pumps are plain bands with a shrink or tight fit in the case. The seal area of the ring to the casing is on the OD of the ring. The OD must be increased to take care of any corrosion or damage to the case if the pump has suffered a seizure-type failure.

Ell-shaped rings are sometimes used in vertically split pumps. They are a good spare part stock item. The seal area is on the vertical face.

8.5.4   Wear ring grooving

Abrasion is a real enemy of wear ring clearance. Surface-hardened materials, hard surface overlay, and heat-treated materials are effective to varying degrees in reducing wear. To further assist, spiral grooves are often used. Apparently small particles of abrasive matter, rotating with the liquid and trying to enter the wear ring clearance, find these grooves and follow them to their outlet. Tests indicate that a 35 percent decrease in leakage can be achieved. Three grooves per inch, $3/32$ in wide by $1/32$-in deep with a right-hand spiral, are commonly used. Experience indicates that the effectiveness is the same regardless of whether the spiral runs counter to or with the flow.

8.5.5   Running clearances

When running clearances between wear rings and between other moving parts, except bearings, are established, consideration must be given to pumping temperatures, suction conditions, the character of the fluids handled, and the expansion and galling characteristics of the ring material. Clearance shall be as liberal as possible. The important consideration is the dependability of operation and freedom from seizure under operating conditions, even at a possible small sacrifice in initial hydraulic efficiency. For cast iron, bronze, hardened 11 to 13 percent chromium, and materials of similar galling tendencies, the clearances found in Fig. 8.5 are typically used for operating temperatures below 500°F. For vertical pumps the running clearances specified in Fig. 8.6 apply to the clearance of steady bearings or interstage bushings when materials of low galling tendency are used.

For cast iron, bronze, hardened 11 to 13% Chromium, and materials of similar galling tendencies, the following running clearances should be used.

Diameter of rotating member at clearance joint, inches	Minimum diametral clearances, inches
Under 2.5	0.011
2.500 to 2.999	0.012
3.000 to 3.499	0.014
3.500 to 3.999	0.016
4.000 to 4.499	0.016
4.500 to 4.999	0.016
5.000 to 5.999	0.017
6.000 to 6.999	0.018
7.000 to 7.999	0.019
8.000 to 8.999	0.020
9.000 to 9.999	0.021
10.000 to 10.999	0.022
11.000 to 11.999	0.023
12.000 to 12.999	0.024
13.000 to 13.999	0.025
14.000 to 14.999	0.026
15.000 to 15.999	0.027
16.000 to 16.999	0.028
17.000 to 17.999	0.029
18.000 to 18.999	0.030
19.000 to 19.999	0.031
20.000 to 20.999	0.032
21.000 to 21.999	0.033
22.000 to 22.999	0.034
23.000 to 23.999	0.035
24.000 to 24.999	0.036
25.000 to 25.999	0.037

NOTES: 1. For materials with severe galling tendencies such as 18-8 stainless steel or operating temperatures above 500°, add 0.005 in to these diametral clearances.

2. There should be a minimum of 50 Brinell hardness *difference* in mating materials.

Figure 8.5 API 610 wear ring clearances.

Vertical pumps using pumpage lubricated sleeve bearings and throat bushings of bronze or carbon running against steel, 416, or 18-8 shafts or sleeves should have the following running clearances.	
Diameter of rotating member, in	Clearance, in
0.75 to 1.50	0.004 to 0.006
1.50 to 2.50	0.006 to 0.008

Figure 8.6   Vertical pump sleeve bearing or bushing clearances.

8.5.6   Wear ring replacement

For most pumps in the medium flow and head range, doubling the wear ring clearance will increase the power requirements about 5 to 7 percent. It is economical to renew a wear ring whenever the initial clearance is doubled.

8.6   Trimming of Pump Impellers

The head and flow developed by a centrifugal pump can be adjusted by trimming the impeller diameter. The resulting performance of the pump is the subject of much confusion. First, what are the effects of the diameter trimming on head, NPSH, flow, efficiency, and vibration? Next, is it best to trim the impeller vanes *and* the shrouds or *just* the vanes?

8.6.1   Hydraulic design considerations

The effects of impeller trimming on pump performance depend very much on the *specific speed* of the impeller. This term classifies the hydraulic features of pump impellers as to type and proportions. It is discussed in detail in Chap. 1. Most refinery pumps fall between about 900 and 2500 on this index. Some vertical multistage pumps range from 4000 to 6000.

8.6.2   Effect on flow and head

With the rotational speed held constant, it is customary to discuss the effects of trimming the impeller by three "affinity laws." These are discussed in detail in Chap. 1. There is a negative discrepancy between the affinity-law-calculated impeller diameter alterations and the actual achieved hydraulic performance of the pump. The affinity-law calculations generally dictate a larger cut than required to effect the desired head and flow reduction. So the impeller should be trimmed to a larger diameter than that calculated. The lower the value of the specific speed of the impeller design and the larger the impeller cut, the larger the discrepancy.

There are several reasons for the actual head and flow being lower than that calculated.

1. The affinity laws assume that the impeller shrouds are parallel. In actuality, the shrouds are parallel only in lower-specific-speed pumps.

2. The liquid exit angle is altered as the impeller is trimmed so the head curve steepens slightly.

3. There is increased turbulent flow at the vane tips as the impeller is trimmed due to increased shroud-to-casing clearance or gap "A."

All these reasons contributed to a reduced head development and flow. Process-type pumps approach mixed flow and are more affected than the true radial flow impellers found in higher-head pumps. Greater caution has to be exercised in altering the diameter of a mixed-flow impeller. The affinity-law errors can be on the order of 20 to 25 percent of the calculated reduction. If the calculated trimming calls for a 10 percent reduction in diameter, only a 7 or 8 percent reduction should be made. If an impeller with a high *specific speed* (an N_s of more than 2000) is to be altered, or if the correction of any impeller amounts to more than about 10 percent, the change should be accomplished in steps. This means first turning down the impeller to a larger diameter than the one calculated and carrying out a trial run. After this test run, the impeller is corrected to its final diameter.

8.6.3 Diameter reduction limits

Impeller diameter reductions for radial designs should be limited to about 70 percent of the maximum diameter or 30 percent reduction. There will be a considerable drop in efficiency, and the head curve may be unstable with greater alterations. For pumps of higher specific-speed values (2500 to 4000), trimming should be limited to about 90 percent of the maximum impeller diameter. The decreased impeller-eye diameter ratio gives less space for the cut.

8.6.4 Effect on efficiency

When pump impeller diameters are reduced, the outlet channel width, blade exit angle, and blade length are altered. If the impeller diameter is reduced appreciably from its maximum design, the geometry is altered greatly and the efficiency may be reduced significantly. Again, the effect of the alteration depends very much on the *specific speed* of the impeller. The greater the impeller diameter reduction from maximum diameter and the higher the specific speed (*not suction specific speed*), the more the pump efficiency will decrease with the trimming of the impeller.

8.6.5 Effect on NPSH

If only a small correction (5 percent or less) to the impeller outside diameter is necessary, it may be assumed that the required NPSH will increase only slightly. Diameter reductions greater than about 5 to 10 percent will increase the NPSH required. This increased NPSH required occurs because specific vane loading is raised by the reduced vane length, affecting the velocity distribution at the impeller inlet. Not all pump companies consistently show the increased $NPSH_r$ with reduced impeller diameters on their pump curves. A great deal of attention must be paid to this factor when the margin between $NPSH_r$ and $NPSH_a$ is very narrow or the $NPSH_r$ for a pump is extremely low.

8.6.6 Effect on axial vibration

The low-frequency axial vibrations that occur due to eddy flows around the impeller because of excessive impeller shroud to casing clearances, gap "A," and suction recirculation are discussed in Sec. 4.1. Flow disturbances related to suction recirculation and cavitation are always present in both diffuser- and volute-type pumps. As the impeller diameters are reduced, the flow distribution across the exit width of the impeller becomes more unstable and the tendency for the high-pressure liquid to return to the low-pressure side and create tip recirculation is greatly increased. Again, the higher-energy-level pumps (above 200 hp and 650 ft of head per stage) are a major concern.

8.6.7 Effects on radial vibration

Careful machining of the volute or diffuser tips to increase gap "B" while maintaining gap "A" has been used for a number of years to reduce the vane passing frequency vibration greatly. The pulsating hydraulic forces acting on the impeller can be reduced by 80 to 85 percent by increasing the radial gap from 1 to 6 percent. There is no loss of overall pump efficiency when the diffuser or volute inlet tips are recessed—contrary to the expectations of many pump designers. Any change that shortened the diffuser or volute channels was believed to reduce efficiency. The observed efficiency improvement results from the reduction of various energy-consuming phenomena: the high noise level, shock, and vibration caused by vane passing frequency as well as the stall generated at the diffuser inlet.

Figure 4.7 gives recommended dimensions from Dr. Makay for the radial gaps of the pump impeller to the casing.

8.6.8 What to trim

No hard and fast guidelines for the mechanical aspects of impeller trimming exist, but there are several pump construction and hydraulic design factors to consider in making the decision of what to trim.

How the impeller is trimmed will greatly influence the hydraulic performance of the pump as well as the vibration levels experienced. An evaluation of hydraulic characteristics is necessary before one can decide how to trim the impeller.

For volute-type pumps, the entire impeller, vanes, and shrouds *may* be cut. However, in some pumps, this will increase the axial vibration and other problems associated with gap "A" (shroud-to-case clearance) due to the uneven flow distribution at the impeller exit area. The double suction impeller-type pump is especially sensitive to problems caused by increasing gap "A," so trimming the entire impeller is not a good choice. It is best to cut the impeller vanes obliquely, as shown in Fig. 8.7, leaving the shrouds unchanged, or to cut the vanes only, as in Fig. 8.8. Trimming the vanes only tends to even out the exit flow pattern and to reduce the recirculation tendencies at the exit area. Gap "A" should be about 0.050 in (radial) for minimum vibration because of the vane passing frequency.

In most diffuser-type pumps, it is best to trim only the vanes, as shown in Fig. 8.8, in order to control tip recirculation and the ill effects of an increased gap "A." This cut yields a more stable head curve because the tendency for tip recirculation and the possibility of suction

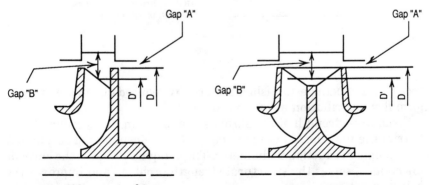

Figure 8.7 Oblique cuts of vanes.

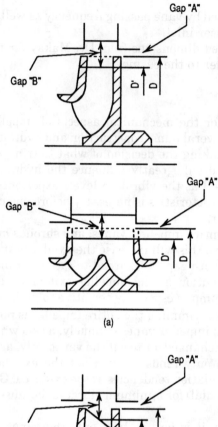

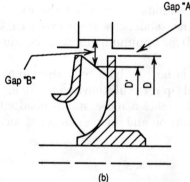

Figure 8.8 Impeller diameter reduction methods: (a) Reduction of shrouds and vanes; (b) reduction of vanes only.

recirculation being established are greatly reduced due to a more uniform flow distribution at the exit area.

Structural strength of the shrouds is a factor in the decision of how to trim the impeller. There may be too much unsupported shroud left after a major reduction in diameter. The oblique cut leaves the shrouds unchanged, solves the structural strength problem, and improves the exit flow pattern.

8.6.9   Correction of vane shapes

Cutting impeller vanes results in blunt vane tips, which cause disturbances in the volute. Impellers manufactured with blunt vane tips can
also cause trouble by generating hydraulic "hammer," even when the
impeller OD is the correct distance from the cutwater (gap "*B*"). The
blunt vane tips cause disturbance in the impeller exit area and in the
volute. Corrections can be made by several methods.

8.6.10   Overfiling

This disturbance may be partly or entirely eliminated by tapering the
vanes by "overfiling" or removing metal on the *leading* face of the
vane, as shown in Fig. 8.9. This technique has the additional advantage of restoring the vane exit angle to nearly that of the maximum
impeller design prior to the diameter reduction.

8.6.11   Underfiling

By sharpening the underside or *trailing* edge of the vane, as shown in
Fig. 8.10, the outlet area of the liquid channel can be enlarged. This
will result in about 5 percent more head near the best efficiency point,
depending on the outlet vane angle. At least ⅛ in of vane tip thickness
must be left. Sharpening the vanes also improves the efficiency

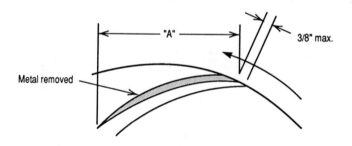

Length of blend for over filing	
Impeller diameter, in	"A" distance of blend, in
10 & below	1½
10¹⁄₁₆ through 15	2½
15¹⁄₁₆ through 20	3½
20¹⁄₁₆ through 30	5
30 & larger	6

Figure 8.9   Impeller vane overfiling.

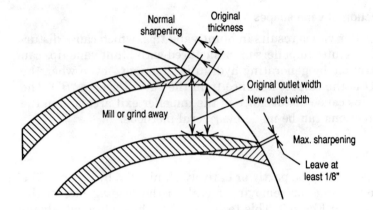

Figure 8.10   Sharpening of impeller vanes by underfiling.

slightly. Where there are high stage pressures, sharpening must be done with great care because of the high static and dynamic stresses on the vanes involved.

Vertical Pumps

9.1 Introduction

Vertical pumps are being used more and more in industry because their operational features are advantageous in many situations such as intermittent storm water, oily-water separators, sewer applications, and cooling-water circulation. Despite the widespread use of this type of pump, many old engineering problems are frequently encountered. The vertical pump's rotor is not gravity-stabilized. There is more cantilever action of the rotor than in a horizontal pump. The gyroscopic effect of rotation can cause lots of damage to the rotor and the casing when problems arise. The control of the liquid flow to the eye of the suction impeller generally is not a function of the pump casing but of a concrete sump.

All parts of the vertical pump are exposed to the fluid being pumped and consequently must be made of suitable material throughout in order to resist corrosive attack. Thus, a vertical process pump made out of higher-alloy materials is considerably more expensive than a horizontal process pump made with the same high alloy and designed for the same service.

Vertical pumps present maintenance problems that are distinctively different from those of horizontally mounted pumps. Many more parts are required to rebuild a vertical turbine pump since typically four stages of this mixed-flow design are required to produce the head of one stage of the radial-flow horizontal pump. More frequent repairs and failures are encountered than with horizontal pumps. This is especially true when the pumped fluid contains some abrasive solids such as sand, scale, or corrosion products.

Disassembly of a vertical pump often is very difficult due to bolting corrosion, shaft abrasion, shaft bowing, and relatively easily breakable cast components. Vibration analysis is almost impossible in vertical pumps. Furthermore, large vertical pumps are expensive to pull and rebuild and entail extended out-of-service time to do this work.

The quality of available pump designs varies widely. Major differences exist in shaft diameter and construction, spacing and length of shaft guide bushings, impeller mounting, casing construction, degree of thrust load (down and/or up), speed selection, and quality of construction. Only top-quality pumps should be considered, particularly for the larger, more critical services. When installed in noncritical processes, vertical pumps are not paid too much attention by engineers in the selection process. Combined with the different maintenance problems that are encountered, this lack of attention can result in increased operating costs.

9.2 Vertical Pump Theory

There are two basic design principles for vertical pumps used in the refining and chemical processing industries:

1. *Radial-flow impeller:* The liquid enters the pump in a plane parallel to the axis of rotation and discharges at 90° to it, as shown in Fig. 9.1. The force or energy imparted to the liquid is virtually all centrifugal. The casing sections can be of a volute or a diffuser design. Normally, the casing is a segmented radial split with bolting on the outer diameter of the casing sections. Radial-flow pumps are characterized hydraulically by relatively higher heads and lower capacities. Radial-flow pumps normally operate at speeds up to 3600 cycles/min or greater.

2. *Vertical turbine impeller:* The liquid enters the pump in a plane parallel to the axis of rotation and is discharged from the impeller at an angle from about 45° to 80° from the axis, as shown in Fig. 9.2. In this design the force or energy imparted to the liquid is a com-

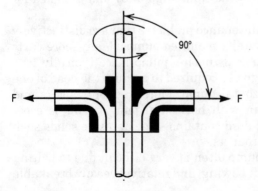

Figure 9.1 Radial-flow impeller —specific speed range 500–3600.

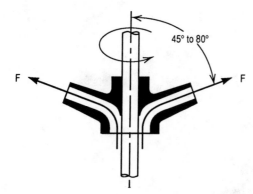

Figure 9.2 Mixed-flow impeller —specific speed range 3600– 8000.

bination of centrifugal force and axial displacement. Casing design is generally of a diffuser section. The basic characteristics of this pump design are as follows: hydraulic head of 10 to 150 ft per stage, a capacity up to 30,000 gal/min, and normal speed of up to 1760 cycles/min.

9.3 Construction Details

Vertical pumps consist of three major components:

1. *Head and driver assembly:* This includes an electric motor and a cast or fabricated base from which the column and casing assembly is suspended. The thrust bearing of the electric motor supports the pump weight and the hydraulic thrust action of the pump in the designs made in the United States. Most European practices incorporate a rolling-element thrust bearing in the pump.

2. *Column and shaft assembly:* The column pipe suspends the pump casing assembly from the head assembly and serves as the conductor for the liquid from the pump casing to the discharge. Within the column pipe is the line shaft, which transmits the power from the driver to the pump impellers. The line shaft bearings or bushings in the column are frequently lubricated by the liquid being pumped. In a few applications an internal oil column is utilized to lubricate the line shaft bearings.

3. *The pump casing assembly:* The pump proper consists of a casing unit or stage with an impeller.

9.4 Vertical Pump Applications

Most vertical process pumps are of the radial-flow "canned" style shown in Figs. 9.3 and 9.4. This design has a very short column or shaft

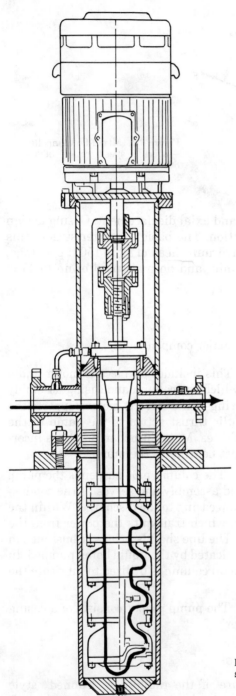

Figure 9.3 Normal casing design—canned type pump.

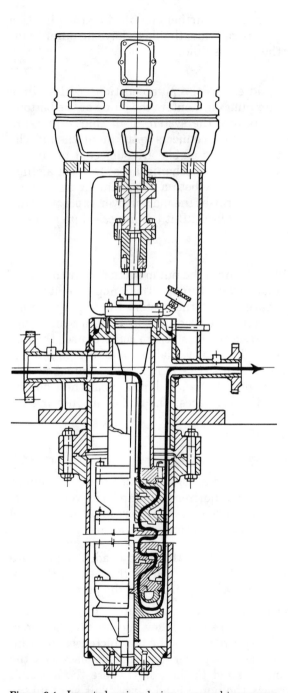

Figure 9.4 Inverted casing design—canned type pump.

assembly. This type of pump can be further classified by the direction of flow through the casing sections. Note that both have an outer "can" or "wrapper" to contain the process fluid.

Normal casing design   Liquid enters the pump suction nozzle, filling the can surrounding the pumping unit, enters the pump at the bottom, is pumped upward, and exits into the system at discharge pressure through the discharge nozzle, which connects to the pump casing. The can and the outside of the pump assembly are subjected to suction pressure only. The design forces the mechanical seal to work against the discharge pressure, which can be both a disadvantage and an advantage. The increased pressure raises the margin over vapor pressure and prevents flashing at the seal faces. But the increased pressure on the seal can be a problem.

Inverted casing design   Liquid enters the pump suction near the top of the pump and is pumped downward through the pump assembly, enters the can, and is discharged into the system through the pump discharge nozzle. The can and the outside of the pump assembly are subjected to the discharge pressure. This design compromises the major reason for using the vertical pump over a horizontal design—lower $NPSH_r$. The mechanical seal of this style pump works against suction pressure. When the pump is used to handle liquids near their boiling point, flashing can occur at the seal faces. The risk of air entering the system is also possible.

9.5 Pump Curve Differences

The performance curves of the radial-flow impeller "canned" pumps approximate those of conventional horizontal pumps. The vertical turbine pump curves are different from those of horizontal pumps and should be reviewed carefully. Two differences are as follows:

1. The shutoff head may be 2 to 3 times the rated head. The pressure capability of the system should be checked for this head if the system flow can be reduced to near-shutoff flow of the pump.
2. The driver horsepower is a maximum near shutoff. The mixed-flow impeller produces axial as well as radial flow. As a result, the horsepower required is relatively constant as the flow increases. In standard centrifugal pumps, the horsepower required increases with the flow. Motors are normally sized to the point where the system's resistance curve intersects the pump head curve. This sizes up the motor at about 110 percent of the power required at the rated flow

point. The motor sizing should be adequate at lower flows if this practice is followed in a vertical pump.

9.6 Vertical Pump NPSH Required

Most pumps are selected to operate at the best efficiency point (BEP). The net positive suction head required (NPSH$_r$) by a pump is usually relatively constant within ±10 percent of the BEP. Vertical turbine pumps generally have a quite rapid rise of the NPSH$_r$ with increasing flow. A vertical turbine pump requiring a 35-ft NPSH at BEP may easily require 60 ft at the end of the usable pump curve. This can be devastating in sump applications where submergence is minimal. The NPSH$_a$ in an open sump is equal to barometric pressure plus submergence minus entry and suction line losses. The NPSH$_a$ can be increased or decreased by increasing or decreasing the sump depth. However, the sump cost is increased as the depth is increased; therefore, the minimum acceptable submergence for the selected pump should be used for the sump design.

In many vertical turbine pump applications, particularly with storm water, the pumps are installed with the discharge going to either an open flume or an unpressurized manifold. The net positive suction head available (NPSH$_a$) of the average sump is about 36 to 37 ft. This is a normal 12-ft-deep sump with 2 to 3 ft of liquid over the centerline of the suction impeller.

9.7 Submergence

The required and available submergence of the suction impeller is a major consideration for vertical turbine pumps. In simple terms, *submergence* is the static elevation difference between the free liquid surface and the eye of the suction impeller. As discussed earlier, submergence and NPSH are not the same. Adequate submergence and low entrance velocities (usually less than 4 ft/s) into the bell mouth are required to avoid vortex formations at the entry of the suction pipe. Inadequate submergence causes random vortices that permit the drawing of air into the pump and, therefore, adverse performance and shortened service life due to damaging vibrations.

Generally, required submergence increases as capacity increases but decreases as speed is decreased. Thus, the selected speed is a major factor in considering the vertical turbine pump. Moreover, the probability of extended trouble-free service increases as the speed is decreased. On the other hand, the cost of both the pump and the driver increases as speed is decreased. Thus, an enlightened judgment must be made for each application regarding operating speed.

9.8 Sump Design

Another factor that can have a serious effect on vertical pump performance is the sump design. The flow of liquid to the pump is under
much less control than in the conventionally piped horizontal pump.
The layout of the sump design frequently causes serious pump malfunctions. Guidelines for pump arrangements; clearance between
pumps, sidewalls, back walls, and sump floor; and perhaps most importantly, the water flow path to each pump are not specific enough
with respect to design details. The fundamental requirement of a good
design is that the liquid not encounter sharp turns or obstructions
which may generate a vortex as it flows to the pump suction bell.
Cavitation, loss of capacity, noisy operation, and high maintenance
costs due to excessive wear will result from the flow disturbances created by steep dropoffs in the floor of the sump. On complex installations with numerous pumps, it is advisable to have the pump manufacturer set up a model test to verify the suitability of the proposed
installation. It is very difficult to modify a sump design after the system has been put in operation. Some simple guidelines for sump design
are as follows:

1. Design the sump for a liquid velocity in the *sump channels* of about
   0.5 to 1 ft/s at the lowest sump water level. The velocity *in the
   channels* should never exceed 1.5 ft/s even under capacity runout at
   reduced-head operation.

2. Avoid sudden changes in the direction of flow. Sump walls, columns
   of adjacent pumps, and channel openings can disturb the flow. The
   pump should be located about one bell mouth diameter off the sump
   wall in all directions.

3. Abrupt changes in sump floor elevations should not be permitted
   within about five bell mouth diameters from the side of the pump.
   The greater the distance, the greater the pump's reliability.

4. The liquid should be flowing parallel to the sump walls when it
   reaches the pump.

5. Avoid columns and cross braces ahead of the pumps whenever possible. If they are necessary, streamline them to avoid the vortex-
   shedding effects.

6. In multiple-pump installations, the intake channel should always
   be perpendicular to the centerline of the row of pumps.

7. The velocity at the *pump suction bell mouth* entrance should be less
   than about 4 ft/s as calculated by the empirical formula[1]

$$V_e = \frac{0.321 \times Q}{0.785 \times D^2}$$

where V_e = velocity in eye of impeller, ft/s
   Q = flow, gal/min
   D = diameter of bell mouth, in

Figures 9.5 and 9.6 show some of these guidelines in practice.

9.9 Vortexing

As discussed in Chap. 3, a vortex is a swirling and funneling action in a liquid. When this occurs in a liquid between the liquid's surface and the draw-off nozzle in a vessel or basin, air or vapor can be drawn into a pump. The formation of a vortex can be very damaging because vortexing effectively reduces the NPSH available. When vortexing occurs between the liquid surface and the draw-off nozzle, air or vapor can be entrained in the liquid and drawn into the pump. With the suction head of a pump close to a minimum value, air entrainment is likely to occur. Only 2 percent air entrainment can result in a 10 percent loss in pump capacity.

The amount of submergence required to prevent the formation of a vortex is a function of the liquid velocity at the bell mouth of the pump.

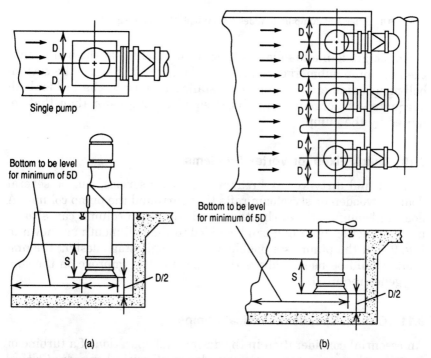

Figure 9.5 Recommended sump designs. (*a*) Single pump installation; (*b*) multiple pump installation.

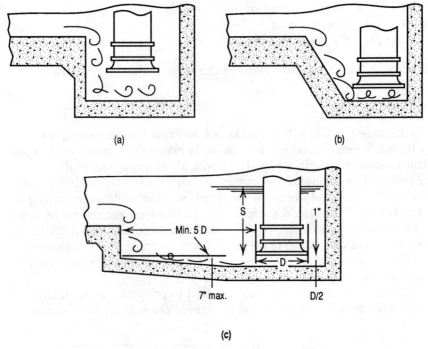

Figure 9.6 Effects of sump bottom design: (*a*) bad; (*b*) bad; (*c*) good.

A simple guideline to follow is this: At about 2 ft/s at the bell mouth, the minimum submergence must be about 1¾ to 2 times the suction bell mouth diameter. A proposed installation must be checked for both adequate submergence and NPSH equal to or greater than that required by the pump.

9.10 Quick Fixes for Vortex Problems

A quick and dirty vortex breaker, for water service, can be several floating wooden or styrofoam rafts that surround the pump column. A floating "doughnut" which fits around the pump column can also be used. A suction umbrella can be added to the bell mouth as shown in Fig. 9.7. If the pump is a double suction single-stage design, the umbrella would be needed above the upper inlet to improve on the submergence required.

9.11 Critical Speeds in Vertical Pumps

An essential consideration in the design and operation of a turbine or centrifugal compressor is the avoidance of critical speeds. Critical

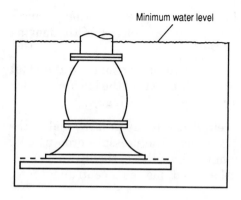

Minimum water level

Figure 9.7 Suction umbrella added to bell mouth.

speeds are usually not a factor in centrifugal pumps because the critical speed bears a close relationship to the presence of liquid in close-clearance spaces of wear rings and interstage bushings. The stiffening and damping effect resulting from the pressure drop across such gaps is large and raises the critical speed far above what is known as the *dry* value.

Critical speeds in pumps are determined by five basic characteristics:

1. Bearing span
   a. Shorter raises
   b Longer lowers
2. Shaft diameter
   a. Larger raises
   b. Smaller lowers
3. Rotor weight—essentially the impeller weights in the bowls and the shaft in the column
   a. Lighter raises
   b. Heavier lowers
4. Liquid film support at leakage points. In a multistage pump this is the greatest influence on critical speed of all the factors.
5. Bearing system damping
   a. Stiffer raises
   b. Looser lowers

It is worth noting that the critical speed of a pump will change slightly with shaft material and when the impeller weight is changed because of a diameter or material change.

A liquid surrounding a pump rotor influences the dynamic behavior in several ways:

• The pressure drop of the liquid in the bushing or wear ring resists shaft displacement.

- The pressure drop of the liquid in the bushing or wear ring produces a restoring force proportional to the lateral velocity of the moving shaft.

- The liquid can develop a restoring force proportional to the shaft acceleration. This is the "hydrodynamic effect" of the bushing, which is like the oil wedge developed in a journal bearing.

The magnitude of the stiffness effect is directly proportional to the pressure drop across the clearance space. The pressure drop usually increases as the square of the pump speed. Hence, the hydrodynamic force increases as the square of the speed, just as a centrifugal force does.

Damping, although it increases the critical speed slightly, is more important for its drastic reduction of deflection if the pump should reach a critical speed.

Thus, hydrodynamic forces work positively for pumps because the wear rings and bushings are closely spaced and numerous in multi-stage pumps. The net effect is a large increase in the critical speed. It is even possible that there will be complete suppression of the critical speed.

A vertical pump has two rotor sections (the pump or bowl section and the shaft column). Because of the hydrodynamic forces, the bowl assembly will have a high critical speed, but the column assembly can have a very low critical speed. Remember, there is no pressure differential across the shaft bearing bushing; the only liquid force influencing the column shaft is a small amount of the liquid wedge discussed in item 3. The column assembly can be near a critical speed because of excess bearing span and low stiffness of the bushings.

9.11.1 Rule of thumb

A rule of thumb for determining critical speeds of vertical pump column assemblies at 1800 r/min is as follows:

$$\text{First critical speed} = \frac{4.76 \times 10^6 \times \text{shaft diameter}}{(\text{bearing span})^2}$$

$$\text{Second critical} = 4 \times \text{first critical}$$

All dimensions are in inches.

9.11.2 Maximum spacing between shaft guide bushings

Through the years some vertical pumps have been manufactured with extremely large spacing between the shaft guide bushings. Two pieces of data are required by the pump applicator to determine if the bushing spacing quoted by the pump vendor is in excess: the pump shaft diameter (in inches) and the pump speed (in revolutions per minute).

The rule of thumb discussed above is in a graph form in API 610 for use in selecting pumps. Problem pumps may be helped by repairs that involve shortening the bearing span. The column spacing between shaft guide bushings shall be 4 ft (1.2 m) or as indicated in Fig. 4 of API 610, 6th edition, paragraph 2.9.2.1, whichever is less. The reasons for this requirement are to reduce the "whip action" of the shaft and to increase the wet critical speeds of the vertical pump. This whip action drastically increases the chance of seal failures and other mechanical problems due to runout on the shaft.

9.12 Thrust Forces

The axial pressure differential forces causing thrust on a vertical turbine pump impeller can be pictured as in Fig. 9.4. Note that both upper and lower impeller shrouds are subjected to discharge pressures, whereas the eye is subjected to only the suction pressure. The real pressures imposed on the various areas of the impeller shrouds are not exactly equal to the discharge pressure. The inequality is due to shroud friction and other factors that are influenced by not only the rotation of the impeller but also the external shape of the impeller and the surrounding bowl structure.

9.12.1 Downthrust—Maximum at shutoff

It is not uncommon to operate a pump against a closed or partially closed valve, resulting in very high downthrust loads. This can occur as part of the start and stop procedure (pumps operating in parallel, pumps on long pipe lines), or it may occur on single units in certain loading and process operations. The thrust at shutoff is very high due to the higher head. This thrust is an axial pressure differential load and can be approximated by

Maximum thrust = constant × shutoff head × specific gravity +

shaft weight + impellers' weight

The term thrust constant can be estimated by the following:

Head (lb/ft) = 0.433(net eye area, in^2)

9.12.2 Upthrust occurs at high flows

In addition to the downthrust force, there is an upthrust force caused by different forces and the inertial load. In the normal operating range of the pump, ever-present upthrust is small in magnitude compared with the downthrust. However, at very high capacities, the actual value in pounds of upthrust force can be very much greater than the downthrust. Thus, the upthrust force becomes a factor when very high capacities occur due to a reaction force caused by the flow. Figure 9.8 shows how upthrust occurs. Radial hydraulic forces A_1 and A_2 cancel. Vertical dynamic or inertial hydraulic forces B_1 and B_2 are additive and unbalanced. As the velocity of flow through a turbine pump impeller increases, the resultant force of the fluid acting on the internal passageways increases in the up direction. A general characteristic of a vertical turbine pump with mixed-flow impellers is that at high flows the upthrust exceeds the downthrust.

The approximate value of this upthrust force F, in pounds, is

$$F = \frac{V_e^2}{2g} \times \frac{\text{net eye area}}{2.31}$$

where V_e = velocity in the eye of the impeller, in feet per second, and the area is in inches squared. The force F increases as the square of the eye velocity, which helps explain why this upthrust force reaches high values at high capacities. If a pump is to be operated at very high capacities on a continuous basis, the resulting upthrust can cause damage in one or more of the following ways:

- Line shafts bend (buckle) due to the compression load and cause vibration and rapid bearing wear.
- Mechanical seals leak due to shaft vibration.
- Mechanical seal malfunctions when the shaft moves upward an excessive amount, thereby changing the fine adjustment between the stationary face and the rotating face of the seal. As a general rule, no seal trouble will be encountered if the vertical shaft of the driver rotating assembly movement is limited to about 0.015 in of movement.

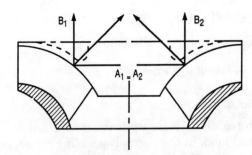

Figure 9.8 Upthrust caused by inertial load on impeller.

- Impellers rub on top of the bowl.
- Driver radial bearings undergo upthrust loads and fail rapidly.
- Driver thrust bearings fail because they can take thrust in only one direction.
- The motor rotor rubs against the stator, causing severe electrical and mechanical damage.
- Ultimate destruction of motor and/or pump may occur due to one or a combination of the above.

The solution to upthrust problems is to start the pump against a pressure on the discharge such as a partially closed discharge valve. As a general rule, no pump should be operated continuously at a capacity greater than 125 percent of full-impeller-diameter peak efficiency capacity, in order to avoid continuous upthrust loads.

9.13 Start-up Considerations

When a pump is first turned on, it usually operates at very high capacities, because the motor reaches speed in just a few seconds and it may take somewhat longer for the head to build up. The pump is therefore likely to operate in the very high capacity range where upthrust occurs. In most installations, the head builds up almost immediately, so that the upthrust is only "momentary." However, even though it is momentary, it is present, and the equipment must be designed to take this upthrust. The pump shaft must run in tension.

Since in many pump applications there is essentially no discharge resistance acting on the pump during start-up, the following sequence occurs that involves loss of suction:

1. The pump immediately goes to over 100 percent of possible flow.
2. The flow velocity through the impeller is at its maximum, and the upthrust rams the motor thrust bearing in the inactive direction. This force can be several times higher than the maximum downthrust.
3. The pump's required head at the high flows exceeds the available head in the sump. The pump experiences a sudden lack of $NPSH_a$, severely cavitates, and stops pumping momentarily.
4. Flow is reduced due to cavitation, so the shaft goes in the downthrust direction.
5. When flow has been reduced, sufficient $NPSH_a$ becomes available, and the pump again goes to 100 percent flow, thereby repeating the cycle.

9.14 Bearing Troubles

The bearings of most vertical pumps are lubricated by the fluid being pumped. This is a distinct disadvantage compared with horizontal centrifugal units. Vertical process pumps require special bearing materials. The bearings can be flushed by a clean fluid to improve bearing life.

The lower bearing is especially vulnerable to poor lubrication because there is almost no pressure differential across the bushing to provide lubrication. A flushing line from the discharge, as shown in Fig. 9.9, will extend the pump life considerably. If the pumped fluid contains abrasive material, a clean fluid from another source may be used, provided the process fluid will not be contaminated by the flushing fluid. A cyclone separator like that used on mechanical seals can be utilized to clean up the pump discharge so that it can be used for flush. The external piping must be added such that each bushing is supplied independently so that excessive wear of one bushing does not rob other bushings of flushing-fluid flow. It may be desirable to add rotameters so that bushing wear can be evaluated by noting the liquid flow. This

Flushing fluid under presure

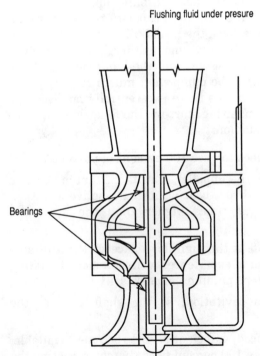

Bearings

Figure 9.9 Vertical pump with bearings flushed.

lubrication design should be used whenever possible on vertical pumps that do not have oil column assemblies.

9.15 Bushing and Bearing Material Selection

The following materials are used as interstage and line-shaft bearing and stuffing box bushings.

1. *Bronze:* Most common material. Do not use on any corrosive service. Do not use above 200 or below −20°F. Do not use on sandy well water. Do not use on dirty-water sump service.

2. *Carbon:* Second choice after bronze. Use when corrosive products preclude the use of bronze and product lubricity is low, e.g., propane containing hydrogen sulfide. Do not use unless pumpage is clean. Use below −20°F. Remember, the coefficient of thermal expansion for carbon is very low compared to that of carbon steel.

3. *Cast iron:* Use on mildly corrosive service, e.g., sour crude. Use for temperatures above 200°F. Use where particulate material in the pumped product precludes the use of carbon. Use only on products with fair or better lubricity.

4. *Cutless rubber:* Applicable only to water pumps. Use in conjunction with open impellers on sandy water or dirty-water sump service. Do not use in hot service or where any product detrimental to neoprene is present.

5. *Teflon, glass-filled or otherwise:* Use only as a last resort and with engineering guidance. Carbon is the best suggestion for most services in which Teflon is specified.

6. *Impregnated graphite:* A solid, self-lubricating, oil-free material consisting of carbon graphite in solid form, the pores of which have been impregnated with molten metal. The metal utilized to fill the pores of the carbon graphite can be babbitt, bronze, iron, nickel, silver, or copper. A carbon-graphite sponge is placed in a vacuum chamber while molten metal at high pressure is driven into and through the sponge, forming long, continuous filaments of metal. These metallic filaments increase the strength and heat-dissipating characteristics of the carbon-graphite base material. The temperature limit of each grade is determined by the metal used for impregnation. The thermal expansion coefficient of bushing material is approximately one-fourth to one-half that of the steel normally used in the "spiders" that support the bushing. The expansion of the

bushing, retainer, and shaft must be considered in calculating the bushing dimensions.

9.16 Bushing Design

The bushing design can be tailored to accommodate various operating problems.

1. A spiral groove is recommended for bushings that have equal pressure at both ends, such as a column bushing. The spiral groove produces a flushing or washing action through the bushing to aid in lubrication. The standard groove is ⅛ in wide by 3/32 in deep with 4-in pitch, right-hand thread.

2. The bushing outside diameter must be supported throughout its entire length. Avoid housing bores that are relieved at midlength.

3. The bushing will close in about 75 to 90 percent of the press fit. The bores must be compensated to allow for this closing.

4. Shoulder the bushing so as to support it against any pressure and/or gravity that will tend to move the bushing out of the housing.

9.17 Vertical Pump Specifications

Vertical pump specifications should have detailed and rigid requirements regarding construction details. These are not well spelled out in many standards. Some of the more important required details include the following:

1. Maximum practical shaft diameter to minimize shaft whip and guide bushing loading.

2. Maximum-length single-piece shafts to avoid eccentricities at shaft section junctures.

3. Corrosion-resistant shafts and internal bolting to facilitate disassembly.

4. Hydraulic axial balance on bowl-impeller assemblies to avoid high thrust loads on the driver bearings.

5. Significant margin in the driver thrust bearing rating to avoid overloading the motor thrust bearing due to axial position of the rotor within the bowl clearances.

6. Dynamically balanced impellers to minimize shaft whip and guide bushing loading.

7. Positive positioning of the impellers on the shaft. Collet mounting should not be permitted except in pumps with very small impellers.

8. Often it is advantageous to look at lower-than-normal speed pumps, particularly for large capacities, to satisfy maximum efficiency, acceptable capacity runout, NPSH requirements, and low submergence. This will increase the pump costs, but these costs will be offset by lower operating costs. Only low-capacity, higher-head pumps should operate at 3600 r/min.

References

1. Johnston Pump Catalog, *Vertical Data*, "Development of an Optimum Sump Design for Propeller and Mixed Flow Pumps."

10

Installation

10.1 General Procedures for the Installation of Centrifugal Pumps

Ensuring the proper installation of centrifugal pumps and their drivers is often overlooked by many end users. The responsibility is normally assigned to installation subcontractors who frequently utilize carpenters and common laborers to do the job. However, the improper installation of rotating equipment leads to increased maintenance costs and premature equipment failure that can cost the industry millions of dollars per year.

The following guidelines apply to the installation of a typical single-stage, horizontal centrifugal pump. The word *contractor* in this text refers to the person responsible for performing the actual work for that installation.

10.1.1 Preinstallation

It is very important that the integrity of the equipment be maintained during the installation phase of a project. Many pieces of equipment have been ruined before they had a chance to operate just due to mishandling in the field prior to a unit start-up. The contractor should inspect the equipment upon arrival and protect it during any on-site storage and construction to prevent corrosion and damage.

All unpainted machined surfaces should be covered with a suitable rust preventive. If more than 6 months will pass before the equipment will be started, consideration should be given to respraying the pump internals with a suitable rust preventive. Ensure that whatever is used is compatible with any elastomers that it may come in contact with, is easily removed, and is applied in accordance with the manufacturer's instructions.

To keep the weather and foreign debris from getting into the pump, all nozzles and openings should be kept covered until the piping is attached. The cost to disassemble a pump just to remove a soda can during a unit start-up can be very expensive.

The contractor should fill all lubricated bearings with the proper lubricating oil as soon as possible. If the bearing is to be oil-mist-lubricated, consider attaching and using the mist generator during the entire construction phase of the project. Greased bearings present a different problem. All greased bearings should be repacked with the correct grease according to the manufacturer's instructions. However, ensure that *all* the old grease is displaced by the new grease. Different greases have different additives that normally are not compatible with each other. Mixing two noncompatible greases will frequently result in a mixture having properties that are less than those of either of the original greases. To prevent rotor sags, the contractor should rotate the shaft of idle equipment approximately 180° every week.

To prevent corrosion of the shaft sleeve, packed pumps received with the packing installed should have the packing removed on arrival and the sleeve and packing gland greased. New packing should be installed just prior to start-up.

Mechanical seals are precision components. On new installations, if a mechanical seal is to fail, normally it will do so within the first few hours of operation. The primary cause of these failures can be traced to improper installation of the seals or flush piping and/or mishandling of the pumps during the installation process. Therefore, pumps with mechanical seals require special handling during transport and installation. During transport, the pumps should be securely restrained to prevent excessive vibration and/or damage to the shaft and seal by dropping or striking. When the pump is moved, do not use the shaft for a leverage or lifting point. Pumps shipped with the mechanical seals installed should be left that way. Locating cams on the seals should be in the preset position. However, vertical pumps may have to have the seals removed in order to make the final alignment between the pump shaft and the driver.

10.1.2 Foundation and anchor bolts

Assuming that the forms and rebar are all sized and placed according to the approved drawings, the next most important step to ensuring a good installation is the placement of the anchor bolts. Prior to making the concrete pour, the following should be accomplished:

- Anchor bolts should be accurately set according to the foundation drawings and firmly secured to prevent shifting during the pouring process.

- Anchor bolts should be dimensionally checked (and rechecked) against the foundation drawing for proper length, diameter, thread length, etc.
- Anchor bolts should be checked for proper projection. (Check for correct elevation as referenced to an established benchmark.)
- To prevent them from sticking to the concrete, make sure that the threads are protected by heavily greasing or waxing with paste wax. The exposed bolts should be covered with plastic wrap, secured firmly with tie wire.
- Plastic or metal anchor bolt sleeves should be used on all anchor bolts (Fig. 10.1). The sleeves are not intended to encourage careless positioning of the anchor bolts but to allow for slight errors in baseplate hole layouts and small shifting of the anchor bolts during the concrete-pouring process.

Baseplates or soleplates should not be placed on the foundation until a minimum of 10 days have elapsed from the time when the foundation was poured if normal concrete was used. High, early strength concrete may be used to permit earlier setting. However, in either case, baseplates and soleplates should not be placed on the foundation until the concrete has had sufficient time to dry and cure so that at least 85 percent of the shrinkage has taken place.

The surface of the foundation should be chipped to remove all laitance and defective or weak concrete by using a chipping hammer.

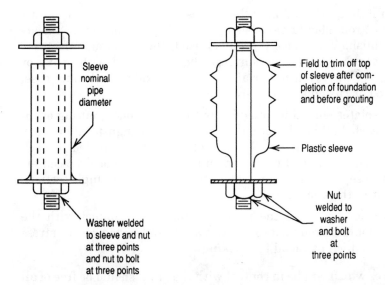

Figure 10.1 Typical anchor bolt sleeves.

Sandblasting or needle gun preparation of the surface is not recommended. The amount of concrete removed should be such that the final baseplate or soleplate elevation results in 1 to 2 in of grout between the surface of the foundation and the lower baseplate flange or the underside of the soleplate. The top surface of the foundation, after chipping, should be rough, reasonably level, and free of all oil, grease, and loose particles. Any oil or grease inadvertently spilled on the surface to be grouted should be removed with a suitable solvent that completely evaporates and leaves no residual film.

Protect the surface of the foundation according to the type of grout to be used. When epoxy grout is used, the concrete surface must be dry when the grout is poured. When cement-based grout is used, keep the foundation surface wet for the time period recommended by the grout manufacturer.

If used, remove the tops of the plastic anchor bolt sleeves and ensure that the sleeves are free of foreign material.

10.1.3 Mounting plate preparation

While the practice varies from one company to another, it is recommended that all pumps and drivers be removed prior to grouting. This aids in leveling the mounting plate and prevents unwanted distortion.

Check all mounting plates (soleplates or baseplates), prior to setting on the foundation to be sure that they are in accordance with the following:

1. To facilitate good bonding, all surfaces that will come into contact with the grout should have been previously sand-blasted to a near-white finish. When epoxy grout is used, the grout manufacturer may recommend that the surfaces then be coated with epoxy or some other approved material, to prevent oxidation. In any case, blistered or rusted surfaces are unacceptable.

2. All baseplates should be provided with at least one grouting opening in each bulkhead section that is large enough to allow easy pouring of the grout. Small vent holes should be drilled in each corner of all bulkhead compartments (Fig. 10.2). For drop-center-trough baseplates, the vent holes should be in the high sections adjacent to the trough.

3. All outside corners of the mounting plates, in contact with the grout, should be rounded to at least a 2-in radius to prevent stress focusing, which can lead to cracking.

All surfaces which will be in contact with the grout should be free of all oil, grease, and rust.

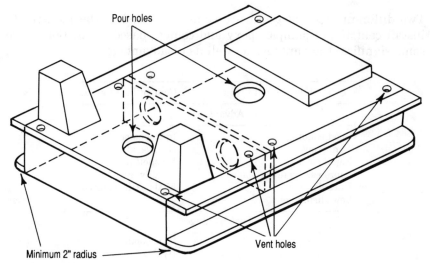

Figure 10.2   Typical pump baseplate.

Position the mounting plate on the prepared foundation. Support it by the leveling screws, leveling shims, or small-taper wedges placed next to or on both sides of the foundation anchor bolts. Cover all leveling screws with a lubricant or wax to prevent the grout from adhering to them. Wax, mask off, or oil all areas which require protection from grout spatter.

Check the elevation of the machined mounting surfaces on the mounting plate.

By adjusting the leveling devices and using a precision level, level the mounting plate to within 0.002 in/ft. It is mandatory that the machined *mounting* surfaces of the mounting plates be horizontal, flat, and parallel. To allow for proper alignment, thus preventing stress and distortion of the equipment, corresponding surfaces in the same plane must be within 0.002 in overall.

When the mounting plate is level, check that all support wedges or shims are in contact with the foundation and plate. Tighten the foundation anchor bolts evenly and recheck the level; adjust as necessary. The mounting plate is now ready for grouting.

10.2   Grouting

10.2.1   General

The role of the grout is to bond the baseplate to the foundation and thereby allow the transmission of any forces and moments, generated from the mounted equipment and piping systems, to the foundation.

Two different types of grout are commonly used in the industry to install centrifugal pumps: epoxy and cement-based grout. Both have some significant advantages as well as disadvantages.

Epoxy grout	Cement-based grout
Advantages	
Good bonding	Inexpensive
Nonporous	No special tools
Rapid-curing	No shelf life
Pours easily; fewer voids	Readily available
Durable	Little skill required
Low shrinkage	
Disadvantages	
High cost	Poor bonding
Shelf life	Porous
Special precautions required	Less flowable
Short pot life	Deteriorates
Watertight forms required	High shrinkage
Skill required	

In the past, because of its cost, the use of epoxy grout was limited to the installation of special-purpose equipment (especially reciprocating machinery) and for the repair of existing foundations. However, in recent years, there has been a trend in the refining industry to use it even for general-purpose equipment such as centrifugal pumps. The justification for the higher initial cost comes from the epoxy grout's higher bonding strength and nonporous finished surface. The increased bond strength helps to maintain the alignment between the driver and the pump, thereby reducing misalignment failures and maintenance costs. The nonporous surface aids in housekeeping and easy cleanup in the event of an accidental spill. With the increased emphasis on environmental concerns, this can be a significant cost saver.

The next two sections provide some guidelines for working with epoxy and cement-based grouts. However, no matter which type of grout is specified, the installation contractor needs to follow the grout manufacturer's written instructions carefully. It is recommended that, for large (either volume or number of) pours, the grout manufacturer have a field representative on site during the installation.

10.2.2 All grout types

Grout forms should be leak-tight and should be securely anchored and shored to withstand the pressure of the grout under working conditions (Fig. 10.3).

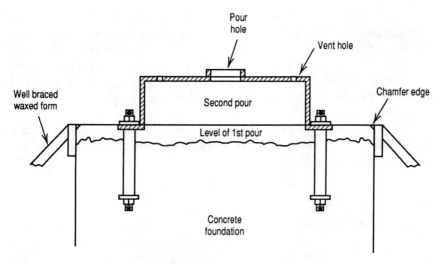

Figure 10.3 Placement of grout forms.

The tops of the grout on baseplates with flange-type support should be at the top of the flange and on soleplates and solid-sided baseplates 1 in above the bottom plane. The outside top edge of the grout should be chamfered at 45°.

Foundation anchor bolt holes should be filled with a nonbonding, pliable material, such as caulking, to prevent a water pocket around the anchor bolt.

A split hose or duct tape should be placed around the exposed threads of the anchor bolts to prevent direct contact between the grout and the bolts.

Timing and proper mixing are the secrets to successful grouting. The grout supplier's instructions must be followed explicitly. Before the components are mixed, everything else should be ready; surfaces cleaned; forms completed and sealed; pouring tools, rags, cleaning solvents, and adequate workforce available.

Generally it is best to start placing the grout at one end of the baseplate and work toward the other end in such a manner as to force the air out from beneath that could cause voids. A funnel or floating tower box should be used to provide a constant head on the grout, which helps the flow under the baseplate. Check the forms frequently for leaks because the leaks are not self-sealing. If not stopped, leaks will cause voids.

The leveling shims and wedges used to level the baseplate should be left in place after grouting. If for some reason they are removed after grouting, the resulting voids should be filled with grout without aggregate. If leveling screws were used, they should be removed and the holes filled with a suitable caulking.

After the grout has cured, the baseplate should be checked for complete grouting by tapping it with a steel bar or hammer. If grouting voids are found based on a hollow sound, holes should be drilled at that location and, by using a grease gun, the voids filled with epoxy grout without aggregate. When pressure injection is employed, use a dial indicator on the baseplate to confirm that the baseplate is not being distorted.

After the grout has properly cured, voids have been filled, and the forms have been removed, the foundation anchor bolts can be re-torqued to the proper values.

10.2.3 Epoxy grout

Epoxy materials should be kept away from open flames, high-heat sources, or sparks. Epoxy grout should be mixed in a well-ventilated area. Workers should wear eye protection, gloves, and protective clothing at all times.

To prevent bonding, forms should be coated with a paste wax on areas that will come into contact with the grout. To permit easy cleanup, wax or cover all surfaces where grout may splash during installation.

The foundation should be protected from rain since it is important that it be clean and dry at the time of grouting.

Grouting materials should be stored at temperatures between 70 and 90°F for at least 24 h before mixing.

The normal ambient temperature for pouring epoxy grouts is between 40 and 90°F. Due to the accelerated rate of curing at high temperatures, shading the foundation from summer sunlight before and after grouting may be required. In hot summer weather, it is sometimes preferable to place the grout during the afternoon so that the initial cure will occur during the cooler evening hours.

For best results in cold weather, a temporary shelter around the perimeter of the baseplate and convection-type heaters may be required to prewarm the equipment and foundation. Do not use radiant heaters or open steam. Radiant heating warms the grout upper surface more than below the surface. The upper grout surface therefore cures in a thermally expanded state and, after dissipation of heat, produces stresses that tend to make the grout curl up, resulting in cracks in the concrete at the foundation corners just below the grout line.

Epoxy grouts can generally be handled with the same methods and tools that are used with flow-grade cement-based grouts. Epoxy grouts can be manually mixed in a wheelbarrow by using mortar mixing hoes, or mixing can be done in small cement mixers. Overmixing and/or violent mixing whips air into the grout and results in a weaker bond.

Epoxy grout is very viscous. However, it will flow and seek its own level, given time.

The curing rate depends on the ambient temperature, thickness of pour, and amount of aggregate used. Lower temperatures and very thin layers require more time. Forms may be removed when the grout is adequately cured. This generally occurs in approximately 12 to 24 h or when the surface becomes firm and not tacky to the touch.

10.2.4 Cement-based grout

Dry packing of cement-based grouts is not acceptable.

Grout forms should be coated with form oil on areas that will come into contact with the grout, to prevent unwanted bonding. To permit easy cleanup, oil or cover all surfaces where grout may splash during installation.

Prior to placing the grout, the top surface of the concrete foundation should be saturated with water for the period of time recommended by the grout manufacturer. Excess surface water and water in the foundation bolt holes should be removed just prior to placing the grout.

The temperature of the baseplate, grout, and foundation should be kept between 40 and 90°F during grouting and for a minimum of 24 h afterward.

The placement of the grout should be rapid and continuous so as to avoid cold joints and voids under the baseplate.

The grout should be cured and the forms left intact according to the grout manufacturer's recommendations.

10.3 Mounting, Piping Up, and Alignment

10.3.1 Installation procedure

Remount the driver and driven equipment on the baseplate. Normally for single-stage, overhung, centerline-mounted pumps, no shims are placed between the mounting flanges and the baseplate pedestals. This is done to ensure that the piping-induced forces and moments are transferred through the pedestal to the baseplate and foundation. However, for all other types of horizontal pumps and drivers, shim packs a minimum of ⅛ in thick should be placed under each mounting foot. The shims should be made of stainless steel, and the number of shims in each pack should be kept to a minimum. Shims should straddle the hold-down bolt and should cover the entire support surface. Stepped or tapered shims should not be used.

Initially align the driver to the pipe-free pump by using the techniques and procedures outlined in this chapter; tighten the hold-down bolts. Check for a soft foot by loosening each hold-down bolt in turn while measuring with a dial indicator the movement between the machine foot and the baseplate. If movement on loosening the nut exceeds approximately 0.001 in at any foot, shim changes should be made and the alignment rechecked.

Pipe up the suction and discharge piping to the pump nozzles.

10.3.2 Flange loadings

Pump casings are designed to accept some amount of flange loading from the attached pipe without distorting the case or adversely affecting the alignment to the driver. In the refining and petrochemical industry, design limits for allowable forces and moments about the flanges are provided in API 610. However, even with established standards, piping loads may be harmful to the pump unless proper engineering and good installation practices are followed. Many mistakes have been made in piping up to pumps, with some resulting in equipment destruction or continuing reliability problems that are difficult to diagnose and resolve. The following general practices should be followed in piping up to centrifugal pumps:

1. Never connect piping to the pump prior to grouting.

2. The piping system should be fabricated by starting at the pump flanges and then working toward the pipe rack.

3. Excessive loads to the pump flanges must be avoided during piping fabrication by using temporary braces and supports as required.

4. To prevent resonant vibrations, long piping runs should be supported at unequal distances. The pipe wave can travel through the pipe hangers if they are at equal distances.

5. Careful and generous use of pipe hangers prevents the transmission of vibration through the structure.

6. Secure anchoring and the generous use of expansion joints and loops in the pipe will ensure excessive loads will not be imposed on the pump.

7. Extreme care in the alignment of piping components must be exercised during initial fabrication to prevent problems later. Many times, a completed, tested, and to all appearances acceptable piping system has required cutting, fitting, welding, and retesting due to lack of caution during initial fabrication as to alignment.

8. Review the piping support design carefully. A good system design will support the piping loads and forces by using spring hangers and bracing, etc., which do not require removal during normal maintenance.

9. The design settings of all pipe hangers should be clearly indicated in the hanger nameplate, preferably stamped, so that time and the weather will not eradicate this important information.

10. After all the fabrication and testing work is complete, loosen piping flanges. Release all lockpins from spring hangers, and remove any temporary bracing previously used. Adjust the system supports as necessary to free the pump from all piping strains.

11. The piping fabrication error that can produce the largest piping strain is nonparallel flange faces. A feeler gage will help to detect this problem. Generally speaking, if you can visually detect a difference in the two facing flange planes, piping strain will result.

12. To check for piping strain, place dial indicators to monitor both vertical and horizontal movement of the pump shaft. Suction and discharge flanges should be made up separately, with indicator readings observed continuously. Should movement exceed 2 mils, piping strain is considered excessive and should be corrected. Loosen pipe until alignment is complete; then tighten and retest for strain.

10.3.3 Installation procedure continued

Recheck alignment and recheck for a soft foot. No significant pipe strain should be present, as would be indicated by an excessive change (greater than 0.002 in) in the pipe-free to piped alignment. Adjust the piping as necessary to eliminate the strain.

After final alignment, the support pads for the pumps should be drilled at two locations for the installation of tapered dowel pins. Unless the manufacturer has recommended a specific location, use Fig. 10.4 as a guideline for placement.

10.4 Preoperational Instructions and Checks and Start-up

10.4.1 Installation procedure continued

All external seal flush and cooling-water piping should be flushed clean prior to connection to the pump.

Prior to operation, the lubricating oil in the bearing housings and constant-level oilers should be drained and new oil added. Verify that the oil level in the bearing housing is correct and that the constant-level oiler is functioning properly.

If the pump is to be lubricated by oil mist, drain the oil from the housing and connect the central mist oil tubing to the housing fittings.

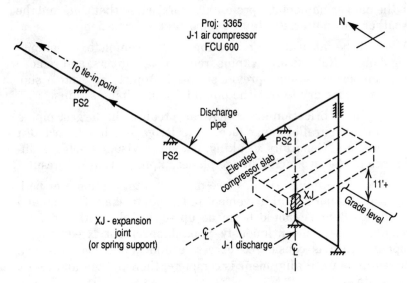

Figure 10.4   Guideline for placement.

Note: The oil mist system should be in operation for a minimum of 12 h prior to an attempt to run any equipment lubricated by the system.

Drivers should be power-rotated to check for proper direction of rotation prior to coupling to the driven equipment. Gear-type couplings mounted on the driver should be provided with a soleplate during the rotational check.

Turbine overspeed trip settings and governor operation should be checked prior to coupling to the driven equipment.

Gear-type couplings should be packed with coupling grease prior to being coupled up. Recheck the coupling float, and ensure that it is within the manufacturer's tolerances.

Flexible disk-type couplings should be installed in accordance with the manufacturer's instructions. Be careful not to damage the disks.

Rotate the coupled train by hand, and if it is free, install the coupling guard.

10.4.2   Start-up

The pump is now ready for the unit start-up. During start-up, minimize the time and temperature used to flush and steam out the pump. This will minimize the amount of debris entering the seal cavity and prevent destruction of the elastomeric seal elements by overheating.

Pumps in high-temperature service (greater than 300°F service and those driven by steam turbines) should have their hot alignment checked. The alignment check should be made as soon as possible after initial operation and as soon as temperatures have stabilized. When the pump is made available, it should be isolated and the coupling guard and spacer removed. Check the hot alignment, and make corrections as necessary. If shims are added, the pump should be thermally cycled again before the alignment is rechecked.

Alignment

11.1 Understanding Alignment

Misalignment accounts for well over one-third of machinery vibration problems. The misalignment of shafting between two pieces of rotating equipment is an operating certainty as well as the most perplexing problem in pump installations. There are three types of misalignment, as shown in Fig. 11.1:

1. Parallel offset misalignment

2. Angular misalignment

3. Combination of parallel offset and angular misalignment

The amount of misalignment can be expressed in degrees, radians, or 0.001-in increments. Although misalignment appears simple, it can be confusing. Some of the confusion arises from a lack of understanding of the coupling's ability to accommodate misalignment. A coupling manufacturer will often quote a misleading value for the amount of misalignment that the coupling will tolerate. One-half degree of an-

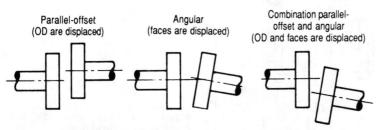

Parallel-offset   Angular   Combination parallel-
(OD are displaced)   (faces are displaced)   offset and angular
                                                   (OD and faces are displaced)

Figure 11.1 Types of misalignment.

gular misalignment is a frequently cited capacity. This amounts to about 8 mils/in of flexing-element spacing, very near lockup. More confusion arises out of a lack of understanding of the thermal growth patterns of the two connected machines. Finally, there are errors in measurement of the relative positions of the connected machines. For smooth operation and high reliability of equipment, knowledge of coupling design, careful evaluation of the thermal rise of the components, and good shaft alignment procedures are essential and cannot be overemphasized.

11.2 Couplings

Two basic types of couplings are used to join pumps and their drivers: *rigid adjustable* and *flexible*.

11.2.1 Rigid adjustable couplings

Most close-coupled vertical pumps utilize a rigid adjustable coupling (Fig. 11.2) to transmit axial load, up or down, to the motor shaft. The pump shaft and the motor shaft are combined into a single shaft by the coupling. The coupling has an adjustable feature in order to control the axial location of the impellers. The proper location of the rotor axially is usually established as a physical measurement of rotor rise from the extreme bottom position. The amount of rise is established by locating the pump shaft nut relative to the motor shaft such that when the coupling is installed and bolted together, the exact axial location of the

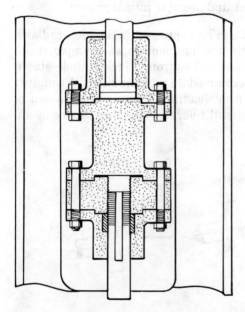

Figure 11.2 Rigid adjustable coupling for vertical pump.

pump shaft is predetermined. The design can transfer both downthrust and upthrust to the motor bearings. Hollow shaft motors are used for deep well turbine type pumps. The pump rotor positioning nut is located on top of the motor. The design cannot transmit upthrust from the pump to the motor.

11.2.2 Flexible couplings

Flexible couplings are used to connect two rotating shafts because changes in temperature and loadings during normal operation, start-up, or shutdown can cause one shaft end to move relative to its companion shaft end. There are two basic types of coupling designs.

1. *Mechanical flexing:* Those couplings which permit misalignment by relative motion between rigid components such as gear teeth, pin and bushing, grid spring, sliding block, etc.

2. *Elastic flexing:* Those couplings which permit misalignment by elastic flexure of an intentionally nonrigid component such as metallic element types or elastomeric designs.

The torque ratings of the coupling assume that the maximum allowable misalignment will be exceeded only briefly as the equipment comes up to normal operating speed. Continuous operation at these limits will cause fatigue failure of the flexing members or rapid wear of the gear teeth on the coupling. In addition, resistance to slipping or flexing by the coupling can cause pump thrust bearing problems. Some gear-type couplings generate as much as 0.33-lb resistance to slipping per transmitted horsepower. This resistance increases as the misalignment increases, thereby increasing thrust bearing loading as the shaft rotates.

Flexing-element separation is a major factor in determining how much misalignment a coupling can tolerate. The greater distance between the coupling hubs permits the coupling to tolerate more angular and offset misalignment. By removing the coupling spacer, sufficient space is created to allow servicing of the mechanical seals of the pump with the driver and pump in place.

11.3 Mechanical Flexing Design Couplings

There are two basic designs of this style of coupling.

11.3.1 Gear couplings

The most common coupling used in the past for horizontal pumps was the flexible gear type (Fig. 11.3). Gear couplings offer high torque and good performance characteristics in relation to their size and weight.

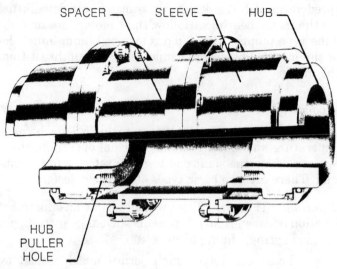

Figure 11.3 Flexible gear coupling.

These couplings consist of five components: two hubs, mounted on the coupled shafts; two sleeve assemblies, which mesh with the hubs; and one floating member, the spacer. The hubs have external teeth with the mating sleeve embodying internal teeth. The relative motion between the meshing gears is in the axial direction and is oscillatory or rocking. This type of motion does not generate a hydrodynamic film of lubrication. Lubrication is provided by a grease or oil which is retained in the gear mesh region by O-rings or other sealing methods. The coefficient of friction on the gear teeth increases as both torque and speed increase, thereby producing moment arms at the pitch diameters which can cause vibration and bearing problems. A major disadvantage of the design is the necessity for lubrication.

11.3.2 Grid spring design

The grid spring coupling consists of two hubs of the coupling which have axial grooves through which the grid spring is laced back and forth. The resilient characteristics of the coupling are achieved through the design of the grooves. A cover keeps the grid spring in place in the grooves and serves as a grease reservoir. It has about 60 percent of the torque transmission capacity per pound of weight of the gear coupling. Again, the need for lubrication is a disadvantage.

11.3.3 Limited end float couplings

The rotor end float in sleeve bearing motors (250 hp up) is from ½ to 1 in. The motor bearing shaft shoulders are inadequate to carry any thrust and must never be in contact when the motor is accelerating, running, or coasting down. The flexible couplings are arranged to restrict the end float of the motor rotor to less than 3⁄16 in, as shown in Fig. 11.4. Any thrust developed within the motor is restricted by the thrust bearing in the driven machine.

The restricting end float is provided for in gear-type or grid-type couplings by locating a "button" at the end of one or both shafts and by inserting a low-friction plate between the two shaft ends. Contact between the hubs and the coupling sleeve prevents excessive movement away from the two shafts.

The rotor end float problem does not arise in ball bearing electric motors, and therefore, they do not require the use of limited end float couplings. Metallic flexing-element type of couplings limit the rotor float by virtue of their design. However, the motor must be set in its running position with no stretch or compression of the coupling flexing element.

11.4 Flexing-Element Couplings

Another form of flexible coupling is the flexing-element coupling. The flexing element can be either elastomeric or metallic.

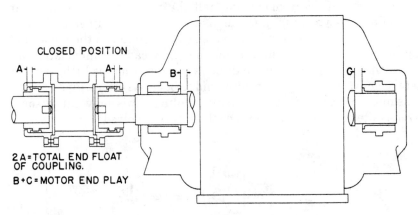

Coupling and bearing settings (Dimensions in inches)					
Nominal motor end play	Limited coupling end float	A	B	C
½	3⁄16	3⁄32	5⁄32	11⁄32
¼	3⁄32	3⁄64	5⁄64	11⁄64

Figure 11.4 Limited end float coupling.

11.4.1 Elastomeric types

The oldest elastomeric design consists of two steel disks, with the driving disk having a number of rigid studs carrying a metal-lined rubber bushing. This projects into corresponding holes in the driven half.

For lower-horsepower ranges, some designs place elastomer blocks in compression while others use the elastomer as a diaphragm. Although the designs eliminate the need for lubrication, they are based on materials of low strength. The size and weight of the designs increase rapidly as torque values go up. Misalignment capabilities are also somewhat reduced compared to the gear coupling.

11.4.2 Metallic types

The coupling consists of two basic designs: the disk (Fig. 11.5) and the diaphragm type (Fig. 11.6). The designs are "soft" in the axial plane so the coupling imposes only about 10 to 15 percent axial bending loads on the bearings or shafts in connected equipment when compared to the gear coupling. This spring rate is nonlinear and does become very stiff at the limits of its stretch. The coupling designs require no lubrication, and the absence of radial clearance throughout the major components means repeatable balance.

In the disadvantage column, the flexing-element coupling designs can accommodate only about one-half the misalignment or angular offset of a gear-type coupling. The *distance between shaft ends* (DBSE) is more critical. These couplings have positive limits on the amount of thermal growth or shaft movement that they can accommodate. Balancing these couplings is more difficult because there is little area for metal removal. The couplings also tend to be heavier for a given torque transmission compared to the gear coupling. This may cause some adverse effects on the dynamics of the rotors involved.

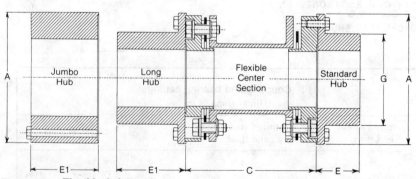

Figure 11.5 Flexible disk coupling.

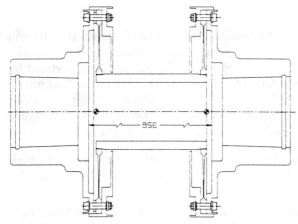

Figure 11.6 Flexible diaphragm coupling.

11.4.3 Disk couplings

The torque is transmitted by the driving bolts pulling the driven bolts through partial arcs of flexible metal disks on identical bolt circles, as shown in Fig. 11.5. The misalignment that the coupling can accommodate is a function of the cube of the arc length of the disk between the bolts. More bolts and shorter flexing arc lengths increase the torque-carrying capabilities but reduce the amount of misalignment that can be tolerated. Frequently the outside diameter of the coupling is increased to improve flexibility. The weight increase can be a serious problem with lightweight rotors.

A "stack" of thin disks is more flexible than one thick disk; therefore most disks range from 0.012 to 0.0015 in in thickness. There are several options in disk shapes, each with advantages and disadvantages:

- *Circular disk:* This shape is the most common because it is the easiest to manufacture. It can tolerate only moderate misalignment and develops high compressive and tensile stresses at the same time in the disk material.

- *Hexagonal disk:* This shape reduces tensile stresses as compared to the circular disk because the line of action of the torque transmission is more centered in the area of the disk. This reduced stress also improves the disk's ability to accommodate the bending stress due to angular misalignment.

- *Scalloped disk:* The tensile stress level over the driving part of the disk is made more uniform by cutting away the least-stressed portion. This greatly increases the flexibility and transfers the maximum bending stresses to the bolt areas.

11.4.4 Diaphragm couplings

Flexibility and torque transmission are achieved between the OD and ID of the diaphragms in this design. A spacer connects the two metallic elements. Misalignment, both angular and axial, is accommodated by elastic flexure of these elements. There is no relative (or "rubbing") movement of any parts of the coupling. The diaphragm elements can be of uniform thickness or can be profiled to improve flexibility, reduce weight, or lower stresses. There are three basic forms:

1. *Tapered contoured:*   The web of the diaphragm is machined (thick at the hub, thin at the rim) to be as flexible as possible from a solid hardened alloy steel disk (Fig. 11.6).

2. *Multiple straight diaphragm with spokes:*   Stresses in a diaphragm tend to increase as the cube of the thickness. To improve flexibility, a number of thin plates are used instead of one thick one.

3. *Multiple convoluted diaphragm:*   This design is very similar to the multiple diaphragm with spokes, except the diaphragm is a nested ripple shape.

11.4.5 Coupling hub specifications

API 671, *Special-Purpose Couplings for Refinery Services,* discusses the coupling-to-shaft juncture. This specification is written for higher-speed compressors and turbines, but it does give pointers for pump couplings. The design of the coupling hub-to-shaft juncture will be essentially the same for all coupling types. This juncture is frequently stressed up to a very high level. If adequate interference is not present, "fretting corrosion" will occur. This subject is discussed later in greater detail.

The following is a summary of general practices in the machinery industry for mounting of coupling hubs.

1. *Straight fit with key (or keys):*   Normally, straight shafts use a very slight interference fit of the hubs on their shafts on the order of 0.0005 to 0.001 in *total* on the shaft diameter. Some maintenance personnel like to bore the coupling to the same clearance on the *loose* side so that the coupling can be easily removed. This is a bad practice because it puts all the torque transmission on the key and causes fretting corrosion of the shaft and coupling hub fit area.

2. *Taper fit with key (or keys):*   The interference on the order of 0.0005 to 0.001 in per inch of shaft diameter is achieved by axially advancing the hub on the shaft. The contact area is about 70 percent. General practice is to mount the counterbored face for the retaining nut about ⅛ in off the shaft shoulder.

3. *Taper fit without keys:*   The interference is greatly increased by use of hydraulic dilation for the hub, and up to 25 percent more torque can be transmitted. API specifications require 0.001 in/in and an 80 percent shaft contact. This interference is generally too light, so most users go to a 0.002 to 0.0025 in/in fit.

11.4.6   Fretting corrosion

The term *fretting corrosion* was coined about fifty years ago, but this complex phenomenon is not fully understood today. The damage done to mechanical joints by fretting corrosion can be major. Coupling hubs and bearing shell-housing fits are the most common systems subject to fretting corrosion. Fretting corrosion not only disrupts surfaces of two mating parts but also destroys tolerances.

The problem of fretting corrosion is complex because it involves both oxidation and wear. The most common explanation of fretting corrosion is that it is mechanical wear which begins by the interlocking action that occurs at contacting high points between two surfaces. These high points are work-hardened and eventually are sheared off by applied tangential forces. A wear particle is produced. As the process continues, the load that was supported by the high spots is more evenly distributed by the accumulated debris. The wear mechanism then shifts from shearing to abrasive action. The oxidation mechanism is a secondary factor. Inspection will reveal a substantial amount of "reddish-brown" debris that forms at the interface of the two components.

Fretting corrosion increases with load, but the important fact seems to be that the coefficient of friction between the two mating parts drops very rapidly. The accumulated debris acts as a shear plane. A coupling hub can become loose and cease to transmit torque. The traditional interference fit of 0.005 in/in of shaft diameter for keyed couplings must be increased to 0.001 in/in to offset this problem.

11.4.7   Keyway design

Coupling keyways should be as specified by AGMA No. 512.01. All keyways should have corner fillet radii, and keys must be chamfered or rounded to clear this fillet radius. The keyway width should be one-fourth the shaft diameter, and the depth should be one-half of the width. One key reduces the shaft strength by about 15 percent from the torsional strength without a keyway. The use of two keys reduces the shaft strength another 7 percent. The cutting of keyways in the hub has about the same effect on its strength. Key-stock should be made of the same material as the shaft.

11.5 Alignment tolerances

The desired target of alignment is line on line or two shafts rotating about the same straight line. These guidelines on alignment apply:

Target: 0.0005 in/in of space between the flexing elements (gear teeth engagement or coupling spacer length)

DBSE: Gear type ±0.030 in of the distance specified; flexing element ±0.010 in

Vibration and other problems should not arise within these tolerances.

11.6 Alignment Methods in Common Use

There are a number of very good methods to measure shaft alignment. A few of the more commonly utilized are as follows:

1. *Sighting in:* Measure the face angles with feeler gages and the coupling hub rim offset with a straightedge.
2. *Rim and face:* Use dial indicators to measure face angles and coupling hub rim offsets, as shown in Fig. 11.7. This is a more accurate version of the sighting-in method, and the principles involved are similar.
3. *Reverse indicator method:* Utilize two sets of brackets and dial indicators, as shown in Fig. 11.8.
4. *Optical alignment:* Use optics to observe shaft positions in both hot and cold conditions.
5. *Laser alignment:* Use of laser beams and the reverse indicator principles achieves a greater accuracy because the indicator spacer length is effectively doubled by the use of reflecting mirrors.

There are enhanced programs based on handheld calculators that make the rim and face and the reverse indicator methods more conve-

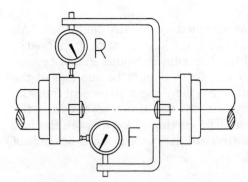

Figure 11.7 Mounting and arranging dial indicators for rim and face method.

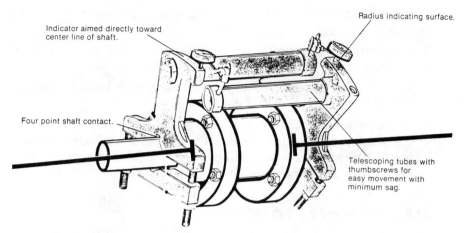

Figure 11.8   Reverse indicator method setup. (*Courtesy A-Line Manufacturng Co.*)

nient, but the basic principles remain the same. The optical and the laser methods can be modified to determine the alignment in both cold (static) and hot (operating) running conditions. Offsetting these advantages is the need for more skilled personnel to operate the equipment for both the laser and the optical methods.

11.7   Details of the Reverse Indicator Method

Although more sophisticated methods of alignment exist, reverse indicator alignment is most commonly used today. The graphic plot used in this method aids greatly in getting the concepts correct. It is a mechanic-oriented method that uses the simplest of tools and the skills of craft personnel to accurately position machinery. The prime advantage is that it presents a picture of the alignment and permits accurate calculations of required shim changes. The rim and face with a feeler gage or the same method with two indicators winds up with an arithmetic approach that is all too easily confused because of plus and minus signs. This confusion frequently results in movement of machines in undesired directions. When coupling hubs are fitted to shaft ends, ideally they should run true, but this is seldom achieved because of manufacturing tolerances between the hub and shaft. To eliminate any eccentricity or runout, it is necessary to rotate both hubs together.

The basic advantages of the reverse indicator method are as follows:

1. Only one measurement surface per shaft is required. The face surface is eliminated.

2. Axial position and movement of the shaft do not influence accuracy. Since the vertical face measurement is eliminated, movement of

either rotor within its thrust bearing clearance does not introduce errors.

3. By turning both coupling halves together, errors due to lack of concentricity and surface irregularities are eliminated.

4. The indicator bracket setup is much more rigid because there is only one indicator mounted per bracket.

5. The reverse indicator method lends itself to the more complicated calculations for multiple-component changes since it gives a physical picture of what is happening to each machine in terms of thermal growth allowances and the needed corrections.

11.8 Understanding Dial Indicators

The individual taking dial indicator readings can easily be confused by several very simple facts. The *total indicator reading* (TIR) is twice the center of shaft offset. Another area of confusion involves understanding the plus and minus signs when dial indicators are used. In the small dial indicators used in alignment, the stem comes out the back of the dial. A typical button-type indicator, the Starrett 196, has a range of 0.200 in and indicates in 0.001-in increments. If the indicator stem is pressed inward toward the dial, the needle sweeps positive (+) and extending the needle sweeps negative (−). The final principle is that an indicator makes a complete sweep, returning to the original value taken off the top of the shaft, and the algebraic sum of the horizontal readings should about equal (within 0.002 in) the algebraic sum of the vertical readings. If they are not equal, something is wrong. One thing that can be wrong is that the centerline of the indicator plunger *must* run through the centerline of the shaft or coupling to give accurate readings. Look at your setup carefully.

11.9 Bracket Sag Measurement

The most misunderstood parameter of alignment is bracket sag. Correction for this sag can mean the difference between a good alignment job and a bad one. The indicator bracket can be custom-fabricated or assembled from a custom-made kit, as shown in Fig. 11.8. It must be sturdy and span the flexible part of the coupling. Sag is always present; the amount varies with the design of the bracket that supports the indicators. The bracket must be built as rigidly as possible, and then the sag must be determined and compensated for in all calculations. No bracket design suffices for all jobs. Many brackets will deflect 0.005 to 0.010 in over a 12- to 15-in span. To check for sag, the bracket is mounted with the indicator in place at the top of a rigid pipe on two solid supports, as shown in Fig. 11.9. The dial indicator is set to read zero. The pipe is rolled 180°, and the number on the dial indicator is read. The difference between the two readings of the dial indicator is the bracket sag. This important number is always a negative value. The sag check should be repeated on the other bracket and indicator.

11.10 Alignment Phases

The following is a brief summary of the planning that should go into the alignment process.

1. *Foundations:*   If the machinery baseplate is not well designed or is not correctly grouted, then good alignment cannot be maintained.

Zero on top

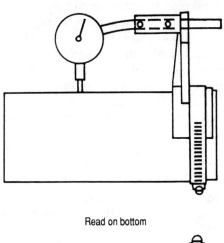

Read on bottom

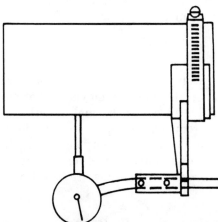

Figure 11.9 Checking for bracket sag.

The most common installation problem encountered is a "soft foot." A soft foot can be located by loosening each hold-down bolt one turn while measuring with a dial indicator any movement between the machine foot and baseplate. If movement exceeds 0.002 in at any foot, shim changes to eliminate the soft foot should be made before proceeding.

2. *Shims:* Shims used are very important to any alignment.
   a. Use only stainless-steel shim stock.
   b. Partial shims should extend on three sides of the hold-down bolts.
   c. Clean all surfaces properly.
   d. Start with a minimum shim package of about 0.125 in under all supports. This is required under API specifications, but frequently is not followed.
3. *Pipe hangers:* Free up and set all pipe hangers correctly.
4. *Doweling:* A simple rule is two dowels per machine in the same transverse plane. Do *not* crisscross dowels, as is frequently encountered in speed reducers. Unless specifically directed by the manufacturer, the dowel pins should be placed as discussed under installation in Chap. 10. The usual dowel pins are no. 2 Morse taper and require 9/16-in holes through the mounting feet, shims, and baseplate. Carefully ream these drilled holes with no. 2 Morse taper reamer, until the dowels fit well and drive tightly with approximately 3/4 in of the dowel protruding. *Note:* Some catalogs may list Morse taper reamers as standard taper reamers to differentiate from other tapers.

Large pumps handling hot liquids are usually doweled near the coupling end, allowing the pump to expand from that end out. Sometimes the other end is provided with a key and a keyway in the casing foot and the baseplate. One method is shown in Fig. 11.10.

11.11 Determination of Targets

The alignment targets should be carefully selected so that they are desired cold locations for the shafts, allowing for elevated thermal growth of the various machines. Allowances for thermal growth can become complicated, since it varies from straight up to straight down or anywhere to the sides. Piping strain can also be very significant in growth patterns. Manufacturers have virtually no feel for the actual heat rise of their equipment. Their recommendations are based on a simple calculation method:

Measure the original vertical height (generally the distance from the machine shaft centerline to the top of the baseplate or soleplate). Estimate the temperature change expected.

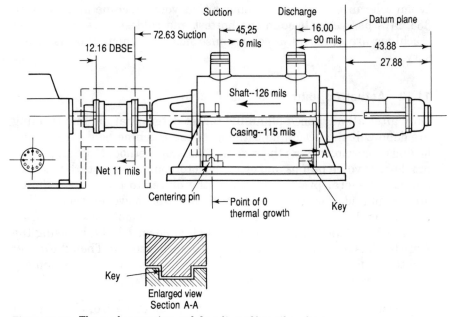

Figure 11.10 Thermal expansion and doweling of hot oil multistage pump.

$$\Delta H = L \text{ (in)} \times 0.0000067 \times \Delta T \text{ (°F)}$$

$$\Delta H = L \text{ (cm)} \times 0.00012 \times \Delta T \text{ (°C)}$$

where ΔH is the change in height and ΔT is the change in temperature.

Normally the vertical movement is minimized by use of a support foot that attaches to the casing centerline. Therefore, the average temperature of the support foot may be considerably lower than that of the adjacent casing. A calculation of this type should be made at each support point of the machine. The results are not totally predictable. Heat rises of 0.040 to 0.050 in greater than those calculated are encountered frequently.

Actual measurement of growth patterns may be done in two distinct steps:

1. Initial alignment is done by using calculated targets from the manufacturer followed by a test run to determine actual thermal growth and other shifting patterns by one of the hot alignment methods: optics, laser, proximity probes, triangular bars, or other methods.

2. The targets should be adjusted for these actual conditions and the machines aligned again. This actual measurement approach is by far the better method for obtaining maximum reliability of machinery.

When tolerances are set, remember that your machine must operate 365 days per year through the hottest and coldest days. Since most machines are outdoors, a compromise heat-rise calculation is necessary.

11.12 Actual Adjustment of the Machine

Only after all advanced planning, target determination, and training are done will the actual adjustment of the machine be done. The following discussion covers the reverse indicator method. Most alignment problems involve pumps and electric motors, pumps and turbines, etc., and it is general practice to adjust the driver to the driven. Piping alignment and other considerations dictate this; however, there will be exceptions. Where a gearbox is used, the alignment *must* start from the gearbox. The gearbox should be bolted down solidly, by using the fewest shims necessary to square it with the baseplate. Then the *driver* and the *driven* unit are aligned to it. The following is a brief summary of the training brochures from A-Line Manufacturing:

1. Measure the machines and lay out the machine from its support centerlines to the other support centerline, as shown in Fig. 11.12. Particular attention should be paid to the DBSE (distance between shaft ends) measurements. Lay out a scale graph with the desired *hot* operating line.

2. Attach the brackets to the shafts, and position the dial indicators on the periphery of the other shaft, as shown in Figs. 11.8 and 11.11. Make sure that the dial indicator anvils are on an axial centerline.

3. Mark both hubs at 90° intervals from the point of contact of the indicator anvils.

4. Rough-align the motor or driver horizontally, following the example in Fig. 11.12. Zero both indicators to the 3 o'clock position. Rotate both machines 180°. Adjust dials to one-half of the readings. Move

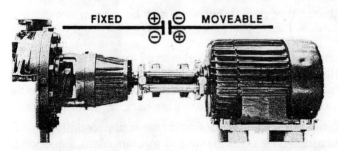

Figure 11.11 Reverse indicator setup. (*Courtesy A-Line Manufacturing Co.*)

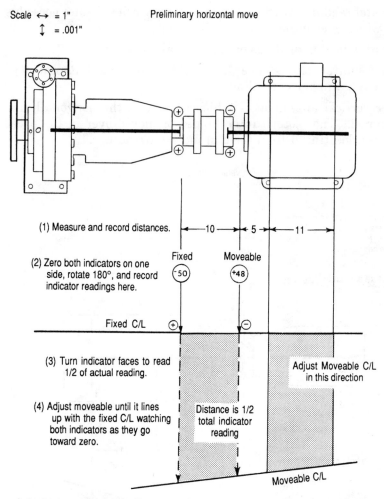

Scale ↔ = 1"
     ↕ = .001"

Preliminary horizontal move

(1) Measure and record distances.

  ←—10—→|←5→|←—11—→

(2) Zero both indicators on one side, rotate 180°, and record indicator readings here.

Fixed   Moveable
(‾50)   (+48)

Fixed C/L ⊕   ⊖

(3) Turn indicator faces to read 1/2 of actual reading.

Adjust Moveable C/L in this direction

(4) Adjust moveable until it lines up with the fixed C/L watching both indicators as they go toward zero.

Distance is 1/2 total indicator reading

Moveable C/L

If alignment cannot be achieved because equipment is boltbound, it is now time to solve the problem.
If proper alignment is achieved, tighten all base bolts

Figure 11.12 Sample problem—horizontal movement. (*Courtesy A-Line Manufacturing Co.*)

the movable machine toward the centerline, while watching both indicators go to zero. Then tighten down both machines firmly. This maintains a more stable and accurate dial indicator reading on the centerline.

5. The vertical alignment is done by setting both indicators at the 12 o'clock position at a plus reading for sag compensation. Slowly rotate both shaft and bracket arrangements together, and record dial

indicator readings at 180° while observing which indicator goes plus and which goes minus.

6. Record indicator readings on a graph, as shown in Fig. 11.13.

7. Add or subtract shims as needed to achieve alignment. Repeat steps 2 through 5 to verify readings.

8. After vertical alignment is achieved, recheck the horizontal misalignment. The motor can be tapped or moved over in the usual manner, locating it from the dial indicator readings.

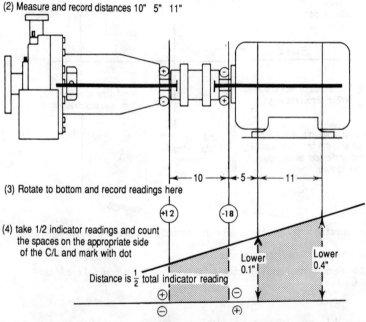

Calculating vertical move

Scale ↔ = 1"

(1) Zero both indicators on top
    or start at a plus reading to
    compensate for sag (if allowing
    for .002" sag, start indicator reading +2

(2) Measure and record distances 10" 5" 11"

←— 10 —→|←5→|←— 11 —→

(3) Rotate to bottom and record readings here

(+12)   (-18)

(4) take 1/2 indicator readings and count
    the spaces on the appropriate side
    of the C/L and mark with dot

Lower
0.1"

Lower
0.4"

Distance is $\frac{1}{2}$ total indicator reading

(+)   (−)

(−)   (+)

(5) With straight edge draw a line connecting   (6) Count the space at the feet of Moveable C/L
    these two dots and extend the line over   To the line on the same plane as the fixed C/L
    the part of the graph representing   is the shim Adjustment.
    the feet

After shim adjustment is satisfactory, go back and complete horizontal move.

IMPORTANT: Be sure to check for correct coupling spacing.

Figure 11.13 Sample problem—vertical movement. (*Courtesy A-Line Manufacturing Co.*)

9. Finally, before tightening down the motor, check the DBSE to the coupling manufacturer's measurement.

10. After the final alignment has been made, the support pads for the driver and the driven should be drilled at two locations, and tapered dowel pins with threaded ends to facilitate removal should be installed.

Provided everything is carried out carefully, readings are taken accurately, and the calculations are correct, the final readings should be within 0.003-in TIR. This misalignment figure is acceptable on most equipment; where a more accurate alignment is required, further manual adjustments may be necessary.

Pump System Troubleshooting

12.1 Introduction

Troubleshooting is the art of applying engineering principles to pump maintenance. Many engineers consider maintenance to be based on industrial mythology or folklore. Therefore good engineering guidelines for determining the causes of many problems are not available. The basic questions of where to start and how to conduct the problem analysis must be addressed.

12.2 Components of a Pumping System

Methodical investigation will help to identify the sources of pump problems. Before you start to troubleshoot the system, take time to review maintenance records. Go to the pump and have the operator explain what was seen. If possible, run the pump and demonstrate the problem. A typical pumping system can be divided into eight component areas for investigation:

1. *The foundation:* Poor foundations, grouting, and flexible baseplate designs can cause many problems.

2. *The driver:* Excitations from the vibrations of the driver (motor, steam turbine, gearing) can be transmitted to other components.

3. *Mechanical power transmission:* Excitations from the coupling area, especially due to misalignment of the driver or eccentrically bored coupling hubs, can be transmitted. Beware of incorrect positioning of the driver and pump such that the distance between shaft ends (DBSE) exceeds the axial flexing limits of the coupling.

4. *The driven pump:*   The design of the pump can greatly influence the hydraulic interaction between the rotor and the casing and thus the problems encountered. Pump thermal-growth misconceptions can create problems.

5. *The suction piping and valves:*   Unfavorable incoming flow conditions such as cavitation, intake vortex, or suction recirculation due to poor design and layout of suction piping and valves can cause flow disturbances.

6. *The discharge piping and valves:*   Unfavorable dynamic behavior of piping because of loads from dynamic, static, or thermal causes, including resonance excitation, can cause trouble.

7. *Instrumentation for control of pump flow:*   Control system–pump interaction during start-ups or other periods of low flow can produce pressure pulsations. High-pressure pulsations can result from the hydraulic instability of the entire pumping system.

8. *Failure to maintain the alignment:*   Once the alignment is established, dowels into the baseplate must hold the pump alignment.

12.3   Vibration-Monitoring Basics

Predictive maintenance in the form of vibration-monitoring trend analysis is important in the hydrocarbon processing industry. The technique is not widely used for determining many potential pump and motor problems. Many technical papers have been published on vibration monitoring to evaluate performance problems on steam turbines and turbocompressors. There are few similar guidelines for pumps. The smaller size and the less critical nature of pumps are factors that cause this disparity. Another factor is that turbines and turbocompressors are pneumatic. The fluid flowing through the machine has relatively little mass compared to the rotor weight. The compressible fluid has only a small impact on the vibration patterns of the rotor. Most of the vibration problems in pneumatic machines are mechanical. A pump, however, is a hydraulic machine. The fluid handled has considerable mass and transmits any pressure pulsations throughout the system undiminished. The interaction between the rotor and casing becomes very important and can be a point of confusion about the vibration patterns of the rotor. Interpretation of vibration data is very difficult.

In every pump, dynamic forces of mechanical and hydraulic origin are present and must be separated and evaluated before solutions can be developed. There are many potential causes of vibration. Some of the causes are shown on the vibration spectrum of Fig. 12.1. The hydraulic forces are as numerous and as great in magnitude as the me-

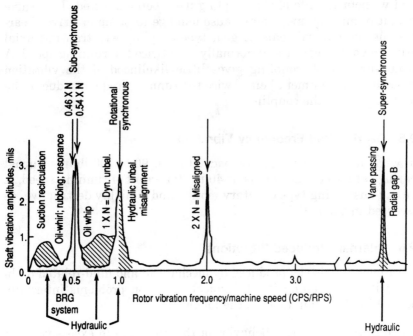

Figure 12.1   Rotor vibration frequency/machine speed—CPS/RPS.

chanical forces. In any analysis program, a distinction between vibrations that are mechanical or hydraulic must be made. A detailed knowledge of the pump's design features as well as its present mechanical condition is necessary to accurately evaluate any vibration data.

12.4   Power Transmission Vibrations

A major source of externally induced vibration is misalignment between the pump and its driver. This misalignment can result in an axial vibration reading as much as 1.5 times the vertical or horizontal readings. Vibration generally occurs at the running speed of the pump, although it may also occur at multiples of the running speed. Vibration caused by misalignment can be distinguished from a resonance disturbance by monitoring the pump during a coast-down period. Misalignment vibration will shift in frequency directly with the rotational speed. Shaft critical speeds or resonances will not shift. Machines that normally operate at elevated temperatures must tolerate vibration during a temporary cold misalignment until the normal operating temperature is reached.

Flexible couplings will accommodate sizable amounts of misalignment without the life of the coupling itself being affected. That same amount of misalignment may cause damage to pump or driver bearings and mechanical seals. A gear-type coupling will transmit axial thrust imposed upon it and normally give twice the running speed. A membrane or disk coupling gives little likelihood of the vibration caused by misalignment being twice the running frequency due to the axial softness of the coupling.

12.5 Bearing Half-Frequency Vibration

Oil whirl may occur in lightly loaded (under 90 to 100 lb/in^2) journal bearings at a frequency of about 40 to 49 percent of running speed. It is a self-sustaining type of rotary motion and is highly destructive, as discussed in Chap. 6.

12.6 Internally Induced Vibrations

Vibrations can also be produced by mechanical and fluid-related problems within the pump itself. Typical vibration problems internally induced by the pump include the following:

1. Unfavorable dynamic behavior of the rotor due to excessive wear ring, bushing, or other leakage clearances.
2. Poor support of the rotor because of loose fits on the shaft or housing in the case of ball bearings. Excessive bore clearance or lack of "clamp" on the shell OD of sleeve bearings causes the same effects.
3. Mechanical imbalance of the rotating parts due to poor balancing or careless assembly. Operational influences including cavitation, erosion, deposits, corrosion, damaged impellers, galled parts, and abrasion can also cause imbalance.
4. Increased axial and radial hydraulic forces when the pump is operated outside of the design flow range. Some increase in vibration is normal when departing from the best efficiency flow rate due to suction recirculation.
5. Pump manufacturer casting and/or machining defects.
6. Radial hydraulic interaction between the rotating-impeller liquid channels and the stationary channels in the casing. This is called the vane passing frequency and is strongly influenced by gap "B."

12.7 Vane Passing Frequency

The vane passing frequency is a hydraulically induced vibration at a frequency determined by the number of impeller vanes, number of stationary vanes, and pump rotational speed. The vibration is created by the momentary disturbance of the wake of the liquid exiting the impeller liquid channels by the stationary diffuser or volute vane tips. gap "*B*" has been discussed in detail in Chaps. 4 and 8. In summary, the larger the gap, the more the flow can smooth out before it contacts a diffuser vane or volute tongue. Determination of the vane passing frequency sounds easy but can be confusing. The most common pump design has an impeller with an odd number of impeller vanes and a double volute in the casing. The vane passing frequency is the rotational speed times the number of vanes. For diffuser-type pumps, the larger number of diffuser vanes coupled with the closer gap "*B*" than that of a volute pump causes a different interaction of the rotor and casing. The match-up of stationary and rotating vanes has a different frequency. This will give a vane passing frequency that does not correspond to the number of vanes in either the impeller or the casing. Figure 12.2 gives a table of common vane interactions from a paper by James Corley, presented at the Fourth Texas A&M Pump Symposium in 1987.

12.8 Random Positioning of Impellers

In the diffuser style pump, a complete set of hydraulics can be created specifically for each pump application because the diffusers are cast separately from the case and the vane angle and location can be readily changed. In the volute design, the volutes can be relocated only by a very expensive pattern change. In multistage volute pumps the degree of positioning of the volutes is severely limited by case design. It is necessary to randomly cut the keyways in the impellers to ensure that vanes on adjacent impellers are not aligned and do not pass volute tongues simultaneously. Frequently, this random keyway positioning is not done when manufacturing the impellers and the vanes on the impellers line up, producing a high vane passing frequency vibration. The alignment of impellers and volutes in each stage should be care-

		Number of impeller vanes			
		4	5	6	7
	8	*	15	9	7
Number	9	8	10	4	28
of	10	6	*	6	21
diffuser	11	12	10	12	21
vanes	12	*	25	*	35
	13	12	25	12	14
	14	6	15	6	3
	15	16	*	6	14

Figure 12.2   Vane passing frequencies for impeller-vane combinations.[1]

fully observed during witness testing, and reassembly on a new pump or when replacing the impellers during maintenance.

12.9 Correction of Vane Tip Shape

Impellers manufactured with blunt vane tips can cause trouble by generating hydraulic disturbances in the impeller exit wake area even when the impeller is the correct distance, gap "*B*," from the cutwater. This disturbance can be greatly reduced by sharpening the impeller vanes on the underside or *trailing* edge of the vane, as discussed in Chap. 8.

12.10 Piping Vibration Limits

How much vibration can be permitted in the piping system of a pump? One rule of thumb states that permissible unfiltered velocity readings taken on the piping at the midspan of its supports can be 3 times the permissible readings taken on the pump bearing caps. Bearing cap readings in the range of 0.5 to 0.6 in/s are the concern level for a pump, and 1.0 in/s is the emergency shutdown level. These pump vibration guidelines then give 1.5 to 3.0 in/s as the limit for piping.

12.11 Mechanical Seal Failures

Troubleshooting mechanical seal failures is a complex task. Vibration analysis does not identify potential seal failures which often can be the deciding factor leading to a pump outage. The number of variables that can affect the seal environment is almost infinite. Many people feel that any failure analysis must be done in a shop situation with the seal dismantled. While this approach is valuable in circumstances where corrosion is present, it cannot cover many factors present in the field. Most efforts to increase the *mean time between failure* (MTBF) have involved installing a more highly engineered seal. Many factors other than seal design, both internal and external to the seal's environment, lead to shorter seal life. The seal must operate in an environment that can be variable and influenced by many factors, some controllable and others uncontrollable. Many are the same factors that influence the hydraulic performance of the pump. Vibration effects caused by poor coupling alignment, shaft runout, bearing condition, and dynamic balance and other mechanical problems are not in the following list.

1. Changes in pump flow rate caused by:
    a. Automatic controllers responding to process demand
    b. Manual adjusting of valves by the pump operators at the pump site

2. Changes in NPSH caused by:
   a. Pressure upsets in the suction vessel
   b. Liquid level changes in towers or drums
   c. Pressure or temperature changes in the process stream
3. Changes in the characteristic of the fluid being sealed due to:
   a. Batch processing
   b. System fluid temperature change due to process or atmospheric changes that alter fluid behavior
   c. Pressure changes which affect the behavior of the fluid
4. External seal flush changes caused by:
   a. Changes in demand on the seal flush system due to bringing other units on- or off-line
   b. Improper manual adjustment of valves in flush lines
   c. Failure of the seal flush system pressure, cooling or heating
5. Changes in flush circulated from the pump to the seal cavity caused by:
   a. Throttle orifice washout or plugging
   b. Flush-fluid heat-exchanger failure
   c. Flush-fluid filter plugging
   d. Cyclone separator flow reversals due to system differential pressure changes
   e. Increased clearance on stuffing box throttle bushings
6. Fluid pressure pulsations (up to 10 percent of head developed) caused by the impeller vane passing frequency and/or suction recirculation at the impeller eye
7. The effects of utility piping for the seal due to:
   a. Changes in the cooling water flow rate to the stuffing box water jackets
   b. Changes in the steam quenching, or steam tracing, of the piping and/or pump

These seven conditions are not always predictable and, in many cases, are just not known. A historical record of operating conditions can be used to correlate mechanical seal performance to increase the mean time between failure. In some applications even the most extreme changes in operating conditions would not be harmful to the mechanical seal life. Other critical applications may require analyzing and controlling these variables in a very precise manner. These factors must be identified and solutions developed for these environmental problems to improve the seal life. These recommended solutions may include a seal design change, pump and process operation changes, or seal fluid environmental changes.

The pump and seal characteristics that have led to repeated seal failures must be understood, and the problems minimized or corrected, so that trouble-free operation can be achieved.

12.12   Evaluation of Pump Operation

The pump operator's descriptions of a problem are rather cryptic: "The pump is not pumping," "the pump is noisy," or "the motor just kicked out." Translating these comments into engineering concepts and values must be done first. Approaching the pump operation problem from a component viewpoint, as outlined in Sec. 12.2, can help in the solution of the problem. Unfortunately, some cavitation and performance problems can be solved only by detailed system calculations. Most troubleshooting guides are oriented more to technical system analysis. The following paragraphs are mental checklists to use at the pump site to observe operating conditions. The list is not complete by any means, but it is a start.

12.12.1   Cavitating-type problems

Problem: Pump is producing a cavitating sound and may or may not be pumping into the system.

Background: Pump does not normally cavitate.

Possible causes	Corrective action
Plugged suction screen	Check for indications of the presence of screen. Remove and clean screen.
Piping gaskets with undersized IDs installed, a very common problem in small pumps.	Install proper-sized gaskets.
Column tray parts or ceramic packing lodged in the impeller eye.	Remove suction piping and debris.
Deteriorated impeller eye due to corrosion.	Replace impeller and overhaul pump.
Flow rate is high enough above design that $NPSH_r$ for flow rate has increased above $NPSH_a$.	Reduce flow rate to that of design.
Lined pipe collapsed at gasket area or ID due to buildup of corrosion products between liner and carbon-steel pipe.	Replace deteriorated piping.
Poor suction piping layout, too many ells in too many planes, a tee branch almost directly feeding the suction of the other pump, or not enough straight run before the suction flange of the pump.	Redesign piping layout, using fewer ells and laterals for tees, and have five or more straight pipe diameters before suction flange.

Possible causes	Corrective action
Vertical pumps experience a vortex formation due to loss of submergence required by the pump. Observe the sump surface while the pump is in operation, if possible.	Review causes of vortexing in Chaps. 3 and 9. Consider installation of a vortex breaker such as a bell mouth umbrella or changes to sump design.
Spare pump begins to cavitate when attempt is made to switch it with the running pump. The spare is "backed off" by running pump because its shutoff head is less than the head produced by the running pump. This is a frequent problem when one pump is turbine driven and one is motor driven.	Throttle discharge of running pump until spare can get in system. Slow down running pump if it is a turbine or variable-speed motor.
Suction piping configuration causes adverse fluid rotation when approaching impeller.	Install sufficient straight run of suction piping, or install vanes in piping to break up prerotation.
Velocity of the liquid is too high as it approaches the impeller eye.	Install larger suction piping or reduce flow through pump.
Pump is operating at a low-flow-producing suction recirculation in the impeller eye. This results in a cavitationlike sound.	Install bypass piping back to suction vessel to increase flow through pump. Remember bypass flow may have to be as high as 50 percent of design flow.

12.12.2 Capacity-type problems

Problem: Pump does not have enough capacity

Background: This is no significant noise from the pump. Experience indicates that the majority of the so-called capacity problems turn out to be a head-limiting problem.

Possible causes	Corrective actions
Check the discharge block valve opening first. It may be paritally closed and thus the problem..	Open block valve completely.
Wear ring clearances are excessive (closed impeller design).	Overhaul pump. Renew wear rings if clearance is about twice design value for energy and performance reasons.
Impeller-to-case or head clearances are excessive (open impeller design).	Reposition impeller to obtain correct clearance.
Air leaks into the system if the pump suction is below atmospheric pressure.	Take actions as needed to eliminate air leaks.

Possible causes	Corrective actions
Increase in piping friction to the discharge vessel due to the following:	Take the following actions:
1. Gate has fallen off the discharge valve stem.	1. Repair or replace gate valve.
2. Spring is broken in the spring-type check valve.	2. Repair valve by replacing spring.
3. Check valve flapper pin is worn, and flapper will not swing open.	3. Overhaul check valve; restore proper clearance to pin and flapper bore.
4. Lined pipe is collapsing.	4. Replace damaged pipe.
5. Control valve stroke improperly set, causing too much pressure drop.	5. Adjust control valve stroke as necessary.
Suction and/or discharge vessel levels are not correct, a problem mostly seen in lower-speed pumps.	Calibrate level controllers as necessary.
Motor is running backward or impeller of double suction design is mounted backward. Discharge pressure developed in both cases is about one-half design value.	Check for proper rotation and mounting of impeller. Reverse motor leads if necessary.
Entrained gas from the process is lowering NPSH available.	Reduce entrained gas in liquid by process changes as needed.
Polymer or scale buildup in discharge nozzle areas.	Shut down pump and remove scale or deposits.
Mechanical seal in suction system under vacuum is leaking air into system, causing pump curve to drop.	Change percentage balance of seal faces or increase spring tension.
The pump may have formed a vortex at high flow rates or low liquid level. Does the vessel have a vortex breaker? Does the incoming flow cause surface to swirl or be agitated?	Reduce flow to design rates. Raise liquid level in suction vessel. Install vortex breaker in suction vessel.
Variable-speed motor running too slowly.	Adjust motor speed as needed.
Bypassing is occurring between volute channels in a double volute pump casing due to a casting defect or extreme erosion.	Overhaul pump; repair eroded area.
Axial positions of impeller(s) are not centered with diffuser vanes. Offset of several impellers will cause vibration and lower head output.	Overhaul pump; reposition individual impellers as needed. Reposition whole rotor by changing thrust collar locator spacer.
When the suction system is under vacuum, the spare pump has difficulty getting into system.	Install a positive-pressure steam (from running pump) to fill the suction line from the block valve through the check valve.

Possible causes	Corrective actions
Certain pump designs use an internal bypass orifice port to alter head-flow curve. High liquid velocities often erode the orifice, causing the pump to go farther out on the pump curve. The system head curve increase corrects the flow back up the curve.	Overhaul pump, restore orifice to correct size.
Replacement impeller is not correct casting pattern; therefore NPSH required is different.	Overhaul pump, replace impeller with correct pattern.
Volute and cutwater area of casing is severely eroded.	Overhaul pump; replace casing or repair by welding. Stress-relieve after welding as needed.

12.12.3   Motor overload problems

Problem: Pump motor kicks out on overload because of high current draw.

Background: No previous electrical system troubles encountered.

Possible causes	Corrective actions
Polymer buildup between wear surfaces (rings or vanes).	Remove buildup to restore clearances.
Excessive wear ring (closed impeller) or cover-case clearance (open impeller).	Replace wear rings or adjust axial clearance of open impeller. In severe cases cover or case must be replaced.
Pump is circulating excessive liquid back to suction through a breakdown bushing or a diffuser gasket area.	Overhaul pump, replacing parts as needed.
Minimum-flow loop left open at normal rates, or bypass around control valve is open.	Close minimum-flow loop or control valve bypass valve.
Discharge piping leaking under liquid level in sump-type design.	Inspect piping for leakage. Replace as needed.
Electrical switch gear problems cause one phase to have low amperage.	Check out switch gear and repair as necessary.
Specific gravity is higher than design specification.	Change process to adjust specific gravity to design value, or throttle pump to reduce horsepower requirements. This will not correct problem with some vertical turbine pumps which have a flat horsepower required curve.

Possible causes	Corrective actions
Pump motor not sized for end of curve operation.	Replace motor with one of larger size, or reduce flow rate.
Open impeller has slight rub on casing. Most often occurs in operations from 250 to 400°F due to piping strain and differential growth in the pump.	Increase clearance of impeller to casing.
A replacement impeller was not trimmed to the correct diameter.	Remove impeller from pump and turn to correct diameter.

12.13 Conclusions

Pump troubleshooting basically consists of paying attention to a lot of details. These details fit together to form an overall picture of the operating problems.

References

1. Corley, James E., "The Vibration Analysis of Pumps, A Tutorial," *Proceedings of Fourth International Pump Symposium,* Texas A & M University, Houston, Tex., May, 1987.

Index

ABOUT THE AUTHORS

John W. Dufour is chief engineer, mechanical equipment services, for Amoco Oil Company in Chicago, and has over twenty years of experience working with mechanical equipment. He is the author of several technical publications, articles, and training programs on mechnical equipment, rotor dynamics, and vibration monitoring.

W. Ed Nelson was manager of maintenance services for Amoco Oil Company in Texas City, Texas until his retirement in 1991. He is the author of more than forty-five technical papers, and a contributor to several handbooks on machinery maintenance and operation. Mr. Nelson also has received six patents on refinery and machinery maintenance techniques.